Monographs Series for Collaborative Innovation on Safety and
Disaster Prevention in Mountainous Urban Construction

MARS Applications in Geotechnical Engineering Systems

Multi-Dimension with Big Data

Monographs Series for Collaborative Innovation on Safety and Disaster Prevention in Mountainous Urban Construction

MARS Applications in Geotechnical Engineering Systems

Multi-Dimension with Big Data

Wengang Zhang

 Science Press
Beijing

 Springer

Wengang Zhang
Chongqing University
Chongqing, China

ISBN 978-981-13-7421-0   ISBN 978-981-13-7422-7   (eBook)
https://doi.org/10.1007/978-981-13-7422-7

Jointly published with Science Press, Beijing, China
The print edition is not for sale in Chinese mainland. Customers from Chinese mainland please order the print book from: Science Press, Beijing, China.
ISBN for the Science Press edition: 978-7-03-061046-1

Library of Congress Control Number: 2019935535

This Springer imprint is published by the registered company Springer Nature Singapore Pte Ltd.
The registered company address is: 152 Beach Road, #21-01/04 Gateway East, Singapore 189721, Singapore

Preface

Many geotechnical engineering problems rely on the use of empirical methods expressed in the form of equations or design charts, to determine the response of the system to input variables. This is usually because of an inadequate understanding of the physical phenomena involved in the multivariate problem, or the system is too complex to be described mathematically. Other surrogate modeling techniques were also developed for implicit geotechnical performance function expressions. Nevertheless, the predictive capacities are less satisfactory for complex multivariate geotechnical problems.

With the rapid increases in processing speed and memory of low-cost computers, it is not surprising that various advanced computational learning tools such as neural networks have been increasingly used for analyzing or modeling highly nonlinear multivariate engineering problems. These algorithms are useful for analyzing many geotechnical problems, particularly those that lack a precise analytical theory or understanding of the phenomena involved. In situations where measured or numerical data are available, neural networks have been shown to offer great promise for mapping the nonlinear interactions (dependency) between the system's inputs and outputs. However, neural networks have been criticized for its long training process since the optimal configuration is not known a priori. This book explores the use of a fairly simple nonparametric regression algorithm known as multivariate adaptive regression splines (MARSs) which has the ability to approximate the relationship between the inputs and outputs, and expresses the relationship mathematically. The main advantages of MARS are its capacity to produce simple, easy-to-interpret models, its ability to estimate the contributions of the input variables, and its computational efficiency. Firstly, the MARS algorithm is described. A number of geotechnical applications with multivariate big data sets are then presented that explore the generalization capabilities and accuracy of this approach.

Chongqing, China   Wengang Zhang

National Joint Engineering Research Center of Geohazards Prevention
in the Reservoir Areas（Chongqing）

Contents

Symbols and Abbreviations

The following list of symbols and abbreviations is provided for ease of reference for those symbols and abbreviations that are most frequently used in this book. It is therefore not exhaustive. Nonetheless, all symbols and abbreviations used in this book are defined at their first mention in the main text.

Symbols

C_U	Coefficient of ununiformity
C_C	Coefficient of curvature
LL	Liqid limit
PI	Plasticity index
G_s	Specific gravity
CP	Collapse potential
ω	Moisture content
γ_d	Dry unit weight
P_w	Pressure at wetting
H_e	Final excavation depth
EI	Wall stiffness
h	Excavation depth at each stage
B	Excavation width
T	Thickness of soft clays
R^2	Coefficient of determination
RMR	Rock mass rating
E_{50}/c_u	Relative soil stiffness ratio
c_u/σ'_v	Relative soil strength ratio
S	System stiffness
δ_{hm}	Maximum lateral wall deflection
μ_w	Water table correction factor
e	Relative error

COV   Coefficient of variation
s_u Undrained shear strength
H_w The length of retaining wall
S_v Average vertical strut spacing
STD   Standard deviation
H   Overburden depth
AR   Advance rate
EP   Earth pressure
E   Soil elastic modulus
GP   Grout pressure
S_t EPB tunneling-induced ground surface settlement
S1   Mean tunnel SPT above crown level
S2   Mean tunnel SPT
r Coefficient of correlation
Q_u Ultimate pile capacity
D_H Lateral ground displacement
R   Nearest horizontal distance to the seismic energy source
M_w Moment magnitude of the earthquake
R^* The modified source distance
T_{15} Cumulative thickness of saturated granular layers
F_{15} The average fines content for granules included within T_{15}
$D50_{15}$ Average mean grain size for granular materials within T_{15}
W   Capacity energy
FC   Fines content
D_r Relative density after consolidation
$CSR_{7.5}$ Cyclic stress ratio
f_s Sleeve friction
σ_v Total vertical stress
a_{max} The peak acceleration at the ground surface
r_d Shear stress reduction factor
MSF   Magnitude scaling factor
z Depth
q_c Measured cone tip resistance
F   Normailized friction ratio
RQD   Rock Quality Designation

Abbreviations

ρ Performance index
ANFIS   Adaptive neuro-fuzzy inference system
ANN   Artificial neural network
ANOVA   Analysis of variance
BF   Basis function

BPF	Blow per foot
BPNN	Back-propagation neural networks
CART	Classification and regression tree
DA	Discriminant analysis
EBBP	Evolutionary Bayesian back-propagation
ff	Free-face ground conditions
GCV	Generalized cross-validation
GP	Genetic programming
gs	Gentle sloping ground conditions
HP-pile	Steel HP-piling piles
LGP	Linear genetic programming
LR	Logistic regression
MAE	Mean average error
MARS	Multivariate adaptive regression splines
MARS_LR	Modified MARS based on logistic regression
MEP	Multi-expression programming
MCS	Maximum compressive stress
MLR	Multiple linear regression
MSE	Mean squared error
MTS	Maximum tensile stress
PR	Polynomial regression
RRMSE	Relative root mean squared error
SPT	Standard penetration test
SVM	Support vector machine

List of Figures

List of Tables

Chapter 1
Introduction

1.1 Background

Many geotechnical engineering problems rely on the use of empirical methods expressed in the form of equations or design charts, to determine the response of the system to input variables, which is generally referred to as the surrogate model or metamodel (metaheuristics). This is usually because of an inadequate understanding of the physical phenomena involved in the multivariate problem, or the system is too complex to be described mathematically. A typical example is the determination of the undrained frictional resistance of piles in clay. Based on field load test data, empirical methods have been proposed in which the adhesion is related to the undrained shear strength as well as other factors such as the pile length by an empirical coefficient.

For problems involving several design (input) variables and nonlinear responses, particularly with statistically dependent input variables, regression methods are usually adopted. Regression methods are well-known mathematical tools for investigating the relationship between dependent variable and independent variable(s) (Montgomery and Peck 1992). Among alternative regression methods, linear regression (LR) is usually preferred in many studies because of its well-established form and available computer packages. This method is based on certain assumptions which must be satisfied for valid results. However, regression models become computationally impractical for problems involving a large number of design variables, particularly when mixed or statistically dependent variables are involved. Another criticism of regression methods lies in their strong model assumptions.

An alternative soft computing technique is the use of artificial neural networks (ANNs). An ANN has a parallel-distributed architecture with a number of interconnected nodes, commonly referred to as neurons. The neurons interact with each other via weighted connections. Each neuron is connected to all the neurons in the next layer. By far, the most commonly used ANN model is known as the back-propagation (BP) algorithm (Rumelhart et al. 1986). In the BP algorithm, the ANN "learns" the complicated model relationship from examples of input and output patterns through

© Science Press and Springer Nature Singapore Pte Ltd. 2019
W. Zhang, *MARS Applications in Geotechnical Engineering Systems*,
https://doi.org/10.1007/978-981-13-7422-7_1

modifying the connection weights to reduce the errors between the actual output values and the target output values. This is carried out by minimizing the defined error function (e.g., sum squared error) using the gradient descent approach. Validation of neural network performance is carried out by "testing" with a separate set of data that was never used in training process, to assess the generalization capability of the trained neural network model to produce the correct input–output mapping.

Generalization is influenced by factors such as the size of the training data, how representative the data are of the problem to be considered, and the physical complexity of the problem. Finding the optimal BP architecture is also important. The BP algorithm has been criticized for its computational inefficiency, i.e., long process to determine the optimal network configuration, since this is not known a priori. Too few hidden neurons may mean that the network is unable to model the nonlinear problem correctly. An excessive number of neurons will result in unnecessary arithmetic calculations and high computation cost and may cause a phenomenon called "overfitting," in which the network learns insignificant aspects of the training set, i.e., the intrinsic noise in the data. Determining the optimal number of hidden neurons is commonly carried out by a trial-and-error approach through repeatedly increasing the number of hidden neurons till no further improvement in the network performance is obtained. Aside from finding the optimal number of hidden neurons and the number of hidden layers, finding the optimal BP architecture is a difficult task that also involves determining the optimal transfer function and learning rate, as well as the maximum number of training cycles (epochs), all of which require considerable computational effort. Various self-pruning NN algorithms have also been proposed, for example initially starting with a network that is a purposely overfit model and then trimming it down to the appropriate size. However, neural networks implemented with these algorithms are generally just as computationally intensive since retraining is required each time a hidden neuron or weighted connection is removed.

As highlighted by Shahin et al. (2008), ANN has been successfully applied to a number of geotechnical engineering problems including pile capacity, settlement of foundations, soil properties and behavior, liquefaction, site characterization, earth retaining structures, dams, blasting and mining, slope stability, geoenvironmental engineering, rock mechanics, tunneling, and underground caverns.

Other commonly used soft computing or machine learning methods include: the support vector machines (Stenwart and Christmann 2008), the Gaussian processes (MacKay 1998), and relevance vector machine (Tipping 2001). The developed predictive models based on these methods are of high accuracy. However, they are generally difficult to interpret.

1.2 Objectives and Scope of This Book

This book presents a nonlinear and nonparametric multivariate adaptive regression spline MARS method to model nonlinear and multidimensional relationships for geotechnical engineering problems. The objectives of this book are as follows:

(1) To introduce a promising MARS procedure for numerical mapping;
(2) To show the main advantages of MARS over other methods for complex data mapping in high-dimensional data;
(3) To present some applications of MARS algorithm in big data geotechnical problems;
(4) To show the procedures of MARS use, including: model development, model interpretation, and parametric sensitivity analysis;
(5) To illustrate the modified MARS procedure for pattern recognition/classification (MARS_LR).

1.3 Outline of This Book

The present book is intended to explore the use of MARS in modeling nonlinear and multidimensional relationships in the following aspects: (a) simple modeling examples; (b) common multivariate geotechnical problems, such as prediction of collapse potential for compacted soils or diaphragm wall deflections in soft clays; (c) big data geotechnical problem such as HP = pile drivability assessment; (d) pattern recognition analysis such as assessment of soil liquefaction or entry-type excavation stability. This book will demonstrate that the use of MARS method provides a promising alternative method for multivariate geotechnical surrogate model building with big data and pattern recognition problems. This book consists of 14 chapters.

This chapter presents the background, objectives, and scope of the study, and an outline of this book.

Chapter 2 gives a condensed review of existing tools and methods for geotechnical surrogate model development. In addition, it introduces the MARS methodology, as well as the main previous applications.

Chapter 3 describes some MARS modeling examples, including the simple function approximation, two-dimensional approximation, function approximation with noise, cowboy hat model approximation.

Chapter 4 presents the MARS use in prediction of collapse potential for compacted soils. MARS model based on a comprehensive database consisting of 192 oedometer tests and 138 similar data sets available in the literature is developed. The developed MARS model, the predictive accuracy, the parametric sensitivity analysis will be provided.

Chapter 5 presents the MARS use in prediction of diaphragm wall deflections in soft clays. MARS model based on a total of 1120 finite element analysis results is built. Comparison between the back-propagation neural network (BPNN) model and MARS model indicates that BPNN gives only slightly better predictions than MARS. However, MARS outperforms BPNN in computational speed and model interpretability.

Chapter 6 presents the MARS use for inverse parameter identification of the soil relative stiffness ratio and the wall system stiffness, to enable designers to determine the appropriate wall size during the preliminary design phase.

Chapter 7 presents the MARS approach for estimating wall deflection profile caused by deep braced excavations, based on an expanded database including a total of 30 case histories for braced excavation in stiff, medium, and soft clays. The developed MARS model can give an accurate graphical representation of the wall deflection profile and estimate the possible depth at which maximum lateral deformation occurs.

Chapter 8 utilizes MARS approach to establish relationships between the maximum surface settlement and the major influencing factors, based on instrumented data on ground deformation and shield operation from three separate EPB tunneling projects in Singapore.

Chapter 9 presents the BPNN and MARS models for assessing HP pile drivability in relation to the prediction of the maximum compressive stresses, maximum tensile stresses, and blow per foot. A database of more than four thousand piles is utilized for model development and comparative performance between BPNN and MARS predictions.

Chapter 10 adopts the MARS method as an improvement to the current MLR model to predict the liquefaction-induced lateral displacement. The predictive accuracy of the developed MARS model, the interpretability, parametric sensitivity analysis, and the design charts derived from the model are presented.

Chapter 11 demonstrates the MARS use for assessment of soil liquefaction via the energy-based approach: capacity energy concept, based on a total of 302 previously published tests. The capacity energies estimated by this proposed model compare favorably with the centrifuge test data sets used for validation purpose.

Chapter 12 presents a modification of the MARS approach based on logistic regression (LR) MARS_LR to evaluate seismic liquefaction potential based on actual field records. Three different MARS_LR models were used to analyze three different field liquefaction databases, and the results are compared with the neural network approaches.

Chapter 13 shows the combined use of MARS approach and LR method MARS_LR for the assessment of rock stability in entry-type excavations, based on an extensive database of cut and fill mining operations and case histories in Canada.

Chapter 14 presents a summary of the findings of this study, with recommendations for future studies.

References

MacKay DJ (1998) Introduction to Gaussian processes. In Neural Networks and Machine Learning, volume 168 of NATO advanced study institute on generalization in neural networks and machine learning. Springer, pp 133–165

Montgomery DC, Peck EA (1992) Introduction to linear regression analysis. Wiley, New York

Rumelhart DE, Hinton GE, Williams RJ (1986) Learning internal representation by error propagation. In: Rumelhart DE, McClelland JL (eds) Parallel distributed processing, vol 1. MIT Press, Cambridge, pp 318–362

Shahin MA, Jaksa MB, Maier HR (2008) State of the art of artificial neural networks in geotechnical engineering. Electron J Geotech Eng 8:1–26

Stenwart I, Christmann A (2008) Support vector machines. Springer, New York

Tipping ME (2001) Sparse Bayesian learning and the relevance vector machine. J Mach Learn Res 1(Jun):211–244

Chapter 2
A Review of Surrogate Models

2.1 Several Commonly Adopted Approaches

One critical aspect before assessing the geotechnical system deformation or stability conditions (responses) is the determination of the limit state surface numerically represented by limit state functions (or performance functions). In many complicated and nonlinear problems where the analyses involve the use of numerical procedures such as the finite element method, this surface may be difficult to determine explicitly in terms of the random variables, and therefore, the limit state function can only be expressed implicitly rather than in a closed-form solution. The following section briefly summarizes the commonly adopted approaches for performance function determination.

2.1.1 Response Surface Methods

The basic concept of RSM is to fit the actual performance function by a closed-form polynomial function with/without cross terms, via selected deterministic analyses and an iterative algorithm. It explores the relationships between several explanatory variables and one or more response variables. This method was introduced by George E. P. Box and K. B. Wilson in 1951. The main idea of RSM is to use a sequence of designed experiments to obtain an optimal response. Box and Wilson suggested using a second-degree polynomial model to do this. They acknowledge that this model is only an approximation, but they use it because such a model is easy to estimate and apply even when little is known about the process. RSM used in geotechnical engineering systems includes Mollon et al. (2009), Lü and Low (2011), Zeng et al. (2014), Lü et al. (2017, 2018).

© Science Press and Springer Nature Singapore Pte Ltd. 2019
W. Zhang, *MARS Applications in Geotechnical Engineering Systems*,
https://doi.org/10.1007/978-981-13-7422-7_2

2.1.2   Regression Analysis

Polynomial regression (PR) and logarithmic regression (LR) models are commonly adopted regression methods to relate the dependent variable to the independent variables and to explore the forms of these relationships. Zhu et al. (2008) predicted the displacements of key points on the side walls of underground openings by using the square and cubic polynomial regressions. Basarir (2008) used LR models to assess the relationship between support pressure, depth, and tunnel deformation for different rock conditions and constructed the response surfaces. Siahmansouri et al. (2012) derived a logarithmic scale formula to predict the ratio of width to height rock pillar in twin circular tunnels. Goh and Zhang (2012) proposed a LR equation to relate the global factor of safety values to tunneling quality index Q and the cavern geometries B and H. Zhang et al. (2015) developed a simple PR model to estimate the maximum wall deflection induced by braced excavation in clays. Zhang and Goh (2015) developed regression models for estimating ultimate and serviceability limit states of underground rock caverns. Goh et al. (2017a, b) proposed a simple regression model for estimation of 3D braced excavation wall deflection. Zhang et al. (2018) established simple regression models relating the excavation-induced pile behaviors to key influential factors and put forward a set of design charts to estimate the behaviors of piles subjected to adjacent braced excavation in soft clay. Gao et al. (2018) developed the logarithmic regression model for investigation of reservoir water level drawdown on slope stability and performed reliability analysis.

2.1.3   Artificial Neural Networks

An ANN has a parallel-distributed architecture with a number of interconnected nodes, commonly referred to as neurons. The neurons interact with each other via weighted connections. Each neuron is connected to all the neurons in the next layer. By far the most commonly used ANN model is known as the back-propagation neural network (BPNN) algorithm (Rumelhart et al. 1986). In the BPNN algorithm, the ANN "learns" the complicated model relationship from examples of input and output patterns through modifying the connection weights to reduce the errors between the actual output values and the target output values. This is carried out by minimizing the defined error function (e.g., sum squared error) using the gradient descent approach. Validation of neural network performance is carried out by "testing" with a separate set of data that was never used in training process, to assess the generalization capability of the trained neural network model to produce the correct input–output mapping. Applications of ANN in geotechnical engineering systems have been reported (e.g., Goh 1994, 1995, 1996; Yang and Zhang 1997; Shi et al. 1998; Juang and Chen 1999; Zhu 2000; Goh 2002; Shahin et al. 2002; Goh and Kulhawy 2003; Shabin et al. 2004; Young and Byung Tak 2006; Baziar and Jafarian 2007; Cao and Qiao 2008; Yilmaz 2009; Pradhan and Lee 2010; Chatzi et al. 2011; Lü et al. 2012;

Goh and Zhang 2012; Zhang and Goh 2013; Conforti et al. 2014; Shooshpasha et al. 2015; Kordnaeij et al. 2015; Zhang and Goh 2016; Cha and Choi 2017; Gao and He 2017). The details of the BPNN algorithm (programmed in MATLAB) adopted in this book are included in Appendix A.

2.2 MARS Methodology

Friedman (1991) introduced MARS as a statistical method for fitting the relationship between a set of input variables and dependent variables.

2.2.1 A Brief Introduction of MARS

MARS is a nonlinear and nonparametric regression method that models the nonlinear responses between the inputs and the output of a system by a series of piecewise linear segments (splines) of differing gradients. No specific assumption about the underlying functional relationship between the input variables and the output is required. The end points of the segments are called knots. A knot marks the end of one region of data and the beginning of another. The resulting piecewise curves (known as basis functions) give greater flexibility to the model, allowing for bends, thresholds, and other departures from linear functions.

MARS generates basis functions by searching in a stepwise manner. An adaptive regression algorithm is used for selecting the knot locations. MARS models are constructed in a two-phase procedure. The forward phase adds functions and finds potential knots to improve the performance, resulting in an overfit model. The backward phase involves pruning the least effective terms. An open-source code on MARS from Jekabsons (2010) is used in carrying out the analyses presented in this book. The details of the MARS algorithm (programmed in MATLAB) adopted in this book are included in Appendix B.

2.2.2 MARS Algorithm

Let y be the target output and $X = (X_1, \ldots, X_P)$ be a matrix of P input variables. Then, it is assumed that the data are generated from an unknown "true" model. In case of a continuous response, this would be

$$y = f(X_1, \ldots, X_P) + e = f(X) + e \tag{2.1}$$

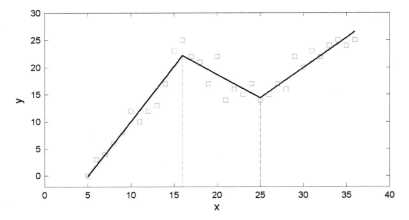

Fig. 2.1 Knots and linear splines for a simple MARS example

in which e is the distribution of the error. MARS approximates the function f by applying basis functions (BFs). BFs are splines (smooth polynomials), including piecewise linear and piecewise cubic functions. For simplicity, only the piecewise linear function is expressed. Piecewise linear functions are of the form $(x - t)_+$ with a knot occurring at value t. The equation $(.)_+$ means that only the positive part of $(.)$ is used; otherwise, it is given a zero value. Formally,

$$(x - t)_+ = \begin{cases} x - t, & \text{if } x \geq t \\ 0, & \text{otherwise} \end{cases} \tag{2.2}$$

The MARS model $f(X)$ is constructed as a linear combination of BFs and their interactions and is expressed as

$$f(X) = \beta_0 + \sum_{m=1}^{M} \beta_m \lambda_m(X) \tag{2.3}$$

where each λ_m is a basis function. It can be a spline function, or the product of two or more spline functions already contained in the model (higher orders can be used when the data warrants it; for simplicity, at most second order is assumed in this book). The coefficients β are constants, estimated using the least-squares method.

Figure 2.1 shows a simple example of how MARS would use piecewise linear spline functions to attempt to fit data. The MARS mathematical equation is expressed as

$$y = 4.4668 + 1.1038 \times \text{BF1} - 3.997 \times \text{BF2} + 1.967 \times \text{BF3} \tag{2.4}$$

where BF1 $= \max(0, x - 16)$, BF2 $= \max(0, 16 - x)$, and BF3 $= \max(0, 25 - x)$ and max is defined as: $\max(a, b)$ is equal to a if $a > b$, else b. The knots are located at $x = 16$ and 25. They delimit three intervals where different linear relationships are identified.

The MARS modeling is a data-driven process. To fit the model in Eq. (2.3), first a forward selection procedure is performed on the training data. A model is constructed with only the intercept, β_0, and the basis pair that produces the largest decrease in the training error is added. Considering a current model with M basis functions, the next pair is added to the model in the form

$$\hat{\beta}_{M+1}\lambda_l(X)(X_j - t)_+ + \hat{\beta}_{M+2}\lambda_l(X)(t - X_j)_+ \tag{2.5}$$

with each β being estimated by the method of least squares. As a basis function is added to the model space, interactions between BFs that are already in the model are also considered. BFs are added until the model reaches some maximum specified number of terms leading to a purposely overfit model.

To reduce the number of terms, a backward deletion sequence follows. The aim of the backward deletion procedure is to find a close to optimal model by removing extraneous variables. The backward pass prunes the model by removing the basis functions with the lowest contribution to the model until it finds the best submodel. Thus, the basis functions maintained in the final optimal model are selected from the set of all candidate basis functions, used in the forward selection step. Model subsets are compared using the less computationally expensive method of generalized cross-validation (GCV). The GCV equation is a goodness-of-fit test that penalizes large numbers of BFs and serves to reduce the chance of overfitting. For the training data with N observations, GCV for a model is calculated as follows (Hastie et al. 2009):

$$\text{GCV} = \frac{\frac{1}{N}\sum_{i=1}^{N}[y_i - f(x_i)]^2}{[1 - \frac{M + d \times (M-1)/2}{N}]^2} \tag{2.6}$$

in which M is the number of BFs, d is the penalizing parameter, N is the number of data sets, and $f(x_i)$ denotes the predicted values of the MARS model. The numerator is the mean squared error of the evaluated model in the training data, penalized by the denominator. The denominator accounts for the increasing variance in the case of increasing model complexity. Note that $(M - 1)/2$ is the number of hinge function knots. The GCV penalizes not only the number of the model's basis functions but also the number of knots. A default value of 3 is assigned to penalizing parameter d (Friedman 1991). At each deletion step, a basis function is removed to minimize Eq. (2.3), until an adequately fitted model is found. MARS is an adaptive procedure because the selection of BFs and the variable knot locations are data-based and specific to the problem at hand.

After the optimal MARS model is determined, by grouping together all the BFs that involve one variable and another grouping of BFs that involve pairwise inter-actions (and even higher-level interactions when applicable), the procedure known as analysis of variance (ANOVA) decomposition (Friedman 1991) can be used to assess the contributions from the input variables and the BFs.

As mentioned previously, the BP algorithm has been criticized for its computational inefficiency, i.e., long process to determine the optimal network configuration since this is not known a priori but has to be determined through a trial-and-error approach. MARS is computationally more efficient at finding the optimal model as it essentially builds flexible models by fitting linear regressions and approximates the model by segmenting separate slopes in distinct intervals of the input variables. The variables to use and the knot locations of the intervals for each variable are determined via a fast but intensive search procedure. The forward selection and backward deletion procedure also ensure that an optimal model can be found.

2.2.3 MARS Applications

Previous applications of MARS algorithm in geotechnical engineering and the relevant fields can be found in various literatures, i.e., model landslide susceptibility mapping (Chen et al. 2018; Conoscenti et al. 2016; Vorpahl et al. 2012; Wang et al. 2015; Felicísimo et al. 2013; Pourghasemi and Rahmati 2017; Pourghasemi and Rossi 2017), prediction of a flexible pavement roughness (Attoh-Okine et al. 2003), investigation of terrain susceptibility to earth-flow occurrence (Conoscenti et al. 2015), reliability analysis for layered soil slopes (Liu and Cheng 2016), prediction of suction caisson uplift capacity (Shahr-Babak et al. 2016; Bhattacharya et al. 2018; Samui et al. 2011), dowel pavement performance modeling (Attoh-Okine et al. 2009), prediction of the lateral load capacity of pile foundation (Samui and Kim 2013; Muduli et al. 2015; Das and Suman 2015), estimation of the lateral spread displacement over a free-face and ground-slope conditions (Liu 2012), determination of the maximum dry density and the unconfined compressive strength of cement stabilized soil (Suman et al. 2016a, b), prediction of the vertical pile capacity of driven pile in cohesionless soil (Mohanty et al. 2016), prediction of the angle of shearing resistance of soil (Samui et al. 2015), determination of horizontal pullout capacity of vertical plate anchors buried in cohesionless soil (Ganesh and Khuntia 2017), predicting tunnel convergence (Adoko et al. 2013), mapping groundwater contamination risk of multiple aquifers (Barzegar et al. 2018), prediction of the maximum magnitude of reservoir-induced earthquakes based on reservoir parameters (Samui and Kim 2012), estimation of raft foundation' settlement under coupled static–dynamic loads (Kaloop et al. 2018), estimations of the compaction parameters of sandy soil (Khuntia et al. 2015), prediction of the elastic modulus of jointed rock mass, prediction of the friction capacity of driven piles in clay (Suman et al. 2016a, b), assessment of scour depth around pipelines (Haghiabi 2016), a quick and accurate measurement of soil characteristics at field (Mohamed et al. 2017), reliability analysis of quick sand condition (Samui et al. 2016), prediction of the spatial variability of reduced level of rock depth (Samui et al. 2015), prediction of the probability of a tree failing during storms (Kabir et al. 2018), system reliability analysis of soil slopes with general slip surfaces (Metya et al. 2017), slope stability analysis (Samui 2013), soil liquefaction assessment (Zhang et al. 2015; Zhang and Goh 2016), reliability assessment of serviceability limit state of twin caverns (Zhang and Goh 2014), pile drivability (Zhang and Goh 2016), prediction of liquefaction-induced lateral spreading (Goh and Zhang 2014), determination of EPB tunnel-related maximum surface settlement (Goh et al. 2018), inverse analysis of soil and wall properties in braced excavation (Zhang et al. 2017a, b), stability evaluation of underground entry-type excavations (Goh et al. 2017a, b), estimation of lateral wall deflection profiles caused by braced excavations in clays (Zhang et al. 2017a, b; Xiang et al. 2018).

2.3 Summary

This chapter briefly summarized the surrogate model building techniques including the RSM, regression analysis, ANN and introduced the MARS algorithm and applications.

References

Adoko AC, Jiao YY, Wu L, Wang H, Wang ZH (2013) Predicting tunnel convergence using multivariate adaptive regression spline and artificial neural network. Tunn Undergr Space Technol 38(3):368–376

Attoh-Okine NO, Mensah S, Nawaiseh M (2003) A new technique for using multivariate adaptive regression splines (mars) in pavement roughness prediction. Transport 156(1):51–56

Attoh-Okine NO, Cooger K, Mensah S (2009) Multivariate adaptive regression (MARS) and hinged hyperplanes (HHP) for doweled pavement performance modeling. J Constr Build Mater 23:3020–3023

Barzegar R, Moghaddam AA, Deo R, Fijani E, Tziritis E (2018) Mapping groundwater contamination risk of multiple aquifers using multi-model ensemble of machine learning algorithms. Sci Total Environ 621:697–712

Basarir H (2008) Analysis of rock-support interaction using numerical and multiple regression modeling. Can Geotech J 45:1–13

Baziar MH, Jafarian Y (2007) Assessment of liquefaction triggering using strain energy concept and ANN model capacity energy. Soil Dyn Earthq Eng 27:1056–1072

Bhattacharya S, Murakonda P, Das S (2018) Prediction of uplift capacity of suction caisson in clay using functional network and multivariate adaptive regression spline. 25(2):1–14

Box GEP, Wilson KB (1951) On the experimental attainment of optimum conditions (with discussion). J Roy Stat Soc B 13(1):1–45

Cao MS, Qiao PZ (2008) Neural network committee-based sensitivity analysis strategy for geotechnical engineering problems. Neural Comput Appl 17:509–519

Cha YJ, Choi W (2017) Deep learning-based crack damage detection using convolutional neural networks. Comput. Aided Civil Infrastruct Eng 32:361–378

Chatzi EN, Hiriyur B, Waisman H, Smyth AW (2011) Experimental application and enhancement of the XFEM–GA algorithm for the detection of flaws in structures. Comput Struct 89(7):556–570

Chen W, Pourghasemi HR, Naghibi SA (2018) Prioritization of landslide conditioning factors and its spatial modeling in shangnan county, China using GIS-based data mining algorithms. Bull Eng Geol Env 77(2):611–629

Conforti M, Pascale S, Robustelli G, Sdao F (2014) Evaluation of prediction capability of the artificial neural networks for mapping landslide susceptibility in the Turbolo River catchment (northern Calabria, Italy). CATENA 113:236–250

Conoscenti C, Ciaccio M, Caraballo-Arias NA, Rotigliano E, Agnesi V (2015) Assessment of susceptibility to earth-flow landslide using logistic regression and multivariate adaptive regression splines: a case of the Belice River Basin (western Sicily, Italy). Geomorphology 242(49):49–64

Conoscenti C, Rotigliano E, Cama M, Caraballo-Arias NA, Lombardo L, Agnesi V (2016) Exploring the effect of absence selection on landslide susceptibility models: a case study in Sicily, Italy. Geomorphology 261:222–235

Das SK, Suman S (2015) Prediction of lateral load capacity of pile in clay using multivariate adaptive regression spline and functional network. Arab J Sci Eng 40(6):1565–1578

Felicísimo ÁM, Cuartero A, Remondo J, Quirós E (2013) Mapping landslide susceptibility with logistic regression, multiple adaptive regression splines, classification and regression trees, and maximum entropy methods: a comparative study. Landslides 10(2):175–189

Friedman JH (1991) Multivariate adaptive regression splines. Ann Stat 19:1–141

Ganesh R, Khuntia S (2017) Estimation of pullout capacity of vertical plate anchors in cohesionless soil using mars. Geotech Geol Eng 2:1–11

Gao W, He TY (2017) Displacement prediction in geotechnical engineering based on evolutionary neural network. Geomech Eng 13:845–860

Gao XC, Liu HL, Zhang WG, Wang W, Wang ZY (2018) Influences of reservoir water level drawdown on slope stability and reliability analysis. Georisk. https://doi.org/10.1080/17499518.2018.1516293

Goh ATC (1994) Seismic liquefaction potential assessed by neural networks. J Geotech Eng 120(9):1467–1480

Goh ATC (1995) Modeling soil correlations using neural networks. J Comput Civil Eng 9:275–278

Goh ATC (1996) Neural-network modeling of CPT seismic liquefaction data. J Geotech Eng 122(1):70–73

Goh ATC (2002) Probabilistic neural network for evaluating seismic liquefaction potential. Can Geotech J 39:219–232

Goh ATC, Kulhawy FH (2003) Neural network approach to model the limit state surface for reliability analysis. Can Geotech J 40:1235–1244

Goh ATC, Zhang WG (2012) Reliability assessment of stability of underground rock caverns. Int J Rock Mech Min Sci 55:157–163

Goh ATC, Zhang WG (2014) An improvement to MLR model for predicting liquefaction-induced lateral spread using multivariate adaptive regression splines. Eng Geol 170:1–10

Goh ATC, Fan Zhang, Zhang WG, Zhang YM, Hanlong Liu (2017a) A simple estimation model for 3D braced excavation wall deflection. Comput Geotech 83:106–113

Goh ATC, Zhang YM, Zhang RH, Zhang WG, Xiao Y (2017b) Evaluating stability of underground entry-type excavations using multivariate adaptive regression splines and logistic regression. Tunn Undergr Space Technol 70:148–154

Goh ATC, Zhang WG, Zhang YM, Xiao Y, Xiang YZ (2018) Determination of EPB tunnel-related maximum surface settlement: a multivariate adaptive regression splines approach. Bull Eng Geol Env 77:489–500

Haghiabi AH (2016) Prediction of river pipeline scour depth using multivariate adaptive regression splines. J Pipeline Syst Eng Pract 8(1):04016015

Hastie T, Tibshirani R, Friedman J (2009) The elements of statistical learning: Data mining, inference and prediction, 2nd edn. Springer

Jekabsons G (2010) VariReg: a software tool for regression modeling using various modeling methods. Riga Technical University. http://www.cs.rtu.lv/jekabsons/

Juang CH, Chen CJ (1999) CPT-based liquefaction evaluation using artificial neural networks. Comput Aided Civ Infrastruct Eng 14(3):221–229

Kabir E, Guikema S, Kane B (2018) Statistical modeling of tree failures during storms. Reliab Eng Syst Saf 177:68–79

Kaloop MR, Hu JW, Elbeltagi E (2018) Pile-raft settlements prediction under coupled static-dynamic loads using four heuristic regression approaches. Shock Vibr. https://doi.org/10.1155/2018/3425461

Khuntia S, Mujtaba H, Patra C, Farooq K, Sivakugan N, Das BM (2015) Prediction of compaction parameters of coarse grained soil using multivariate adaptive regression splines (mars). Int J Geotech Eng 9(1):79–88

Kordnaeij A, Kalantary F, Kordtabar B et al (2015) Prediction of recompression index using GMDH-type neural network based on geotechnical soil properties. Soils Found 55:1335–1345

Liu Z (2012) Prediction of lateral spread displacement: data-driven approaches. Bull Earthq Eng 10(5):1431–1454

Liu LL, Cheng YM (2016) Efficient system reliability analysis of soil slopes using multivariate adaptive regression splines-based monte carlo simulation. Comput Geotech 79:41–54

Lü Q, Low BK (2011) Probabilistic analysis of underground rock excavations using response surface method and SORM. Comput Geotech 38(8):1008–1021

Lü Q, Chan CL, Low BK (2012) Probabilistic evaluation of ground-support interaction for deep rock excavation using artificial neural network and uniform design. Tunn Undergr Space Technol 32:1–18

Lü Q, Xiao ZP, Ji J, Zheng J (2017) Reliability based design optimization for a rock tunnel support system with multiple failure modes using response surface method. Tunn Undergr Space Technol 70:1–10

Lü Q, Xiao ZP, Zheng J, Shang YQ (2018) Probabilistic assessment of tunnel convergence considering spatial variability in rock mass properties using interpolated autocorrelation and response surface method. Geoscience Frontiers. Online https://doi.org/10.1016/j.gsf.2017.08.007

Metya S, Mukhopadhyay T, Adhikari S, Bhattacharya G (2017) System reliability analysis of soil slopes with general slip surfaces using multivariate adaptive regression splines. Comput Geotech 87:212–228

Mohamed ES, Saleh AM, Belal AB, Gad A (2017) Application of near-infrared reflectance for quantitative assessment of soil properties. Egypt J Remote Sens Space Sci 21(1)

Mohanty R, Suman S, Das SK (2016) Prediction of vertical pile capacity of driven pile in cohesionless soil using artificial intelligence techniques. Int J Geotech Eng 12:209–216

Mollon G, Dias D, Soubra AH (2009) Probabilistic analysis of circular tunnels in homogeneous soil using response surface methodology. J Geotech Geoenviron Eng, ASCE 135(9):1314–1325

Muduli PK, Das MR, Das SK, Senapati S (2015) Lateral load capacity of piles in clay using genetic programming and multivariate adaptive regression spline. Indian Geotech J 45(3):349–359

Pourghasemi HR, Rahmati O (2017) Prediction of the landslide susceptibility: which algorithm, which precision? Catena. https://doi.org/10.1016/j.catena.2017.11.022

Pourghasemi HR, Rossi M (2017) Landslide susceptibility modeling in a landslide prone area in Mazandarn Province, north of Iran: a comparison between glm, gam, mars, and m-ahp methods. Theoret Appl Climatol 130(1–2):1–25

Pradhan B, Lee S (2010) Landslide susceptibility assessment and factor effect analysis: back propagation artificial neural networks and their comparison with frequency ratio and bivariate logistic regression modelling. Environ Model Softw 25:747–759

Rumelhart DE, Hinton GE, Williams RJ (1986) Learning internal representation by error propagation. In Rumelhart DE, McClelland JL (eds) Parallel distributed processing, vol 1, MIT Press, Cambridge, pp 318–362

Samui P (2013) Multivariate adaptive regression spline (mars) for prediction of elastic modulus of jointed rock mass. Geotech Geol Eng 31(1):249–253

Samui P, Das S, Kim D (2011) Uplift capacity of suction caisson in clay using multivariate adaptive regression spline. Ocean Eng 38(17):2123–2127

Samui P, Kim D (2012) Modelling of reservoir-induced earthquakes: a multivariate adaptive regression spline. J Geophys Eng 9(5):494–497

Samui P, Kim D (2013) Least square support vector machine and multivariate adaptive regression spline for modeling lateral load capacity of piles. Neural Comput Appl 23(3–4):1123–1127

Samui P, Kim D, Viswanathan R (2015) Spatial variability of rock depth using adaptive neuro-fuzzy inference system (anfis) and multivariate adaptive regression spline (mars). Environ Earth Scie 73(8):4265–4272

Samui P, Kurup P, Dhivya S, Jagan J (2016) Reliability analysis of quick sand condition. Geotech Geol Eng 34(2):579–584

Shabin MA, Maier HR, Jaksa MB (2004) Data division for developing neural networks applied to geotechnical engineering. J Comput Civil Eng 18:105–114

Shahin MA, Maier HR, Jaksa MB (2002) Predicting settlement of shallow foundations using neural networks. J Geotech Geoenviron Eng 128:785–793

Shahr-Babak MM, Khanjani MJ, Qaderi K (2016) Uplift capacity prediction of suction caisson in clay using a hybrid intelligence method (gmdh-hs). Appl Ocean Res 59:408–416

Shi J, Ortigao JAR, Bai J (1998) Modular neural networks for predicting settlements during tunneling. J Geotech Geoenviron Eng, ASCE 124(5):389–395

Shooshpasha I, Amiri I, MolaAbasi H (2015) An investigation of friction angle correlation with geotechnical properties for granular soils using GMDH type neural networks. Scientia Iranica 22:157–164

Siahmansouri A, Gholamnejad J, Marji MF (2012) A new method to predict ratio of width to height rock pillar in twin circular tunnels. J Geol Geosci 1:103. 1:1, https://doi.org/10.4172/jgg.1000103

Suman S, Das SK, Mohanty R (2016a) Prediction of friction capacity of driven piles in clay using artificial intelligence techniques. Int J Geotech Eng 10(5):469–475

Suman S, Mahamaya M, Das SK (2016b) Prediction of maximum dry density and unconfined compressive strength of cement stabilised soil using artificial intelligence techniques. Int J Geosynthetics Ground Eng 2:11

Vorpahl P, Elsenbeer H, Märker M, Schröder B (2012) How can statistical models help to determine driving factors of landslides? Ecol Model 239(1):27–39

Wang LJ, Guo M, Sawada K, Lin J, Zhang J (2015) Landslide susceptibility mapping in Mizunami City, Japan: a comparison between logistic regression, bivariate statistical analysis and multivariate adaptive regression spline models. CATENA 135:271–282

Xiang YZ, Goh ATC, Zhang WG, Zhang RH (2018) A multivariate adaptive regression splines model for estimation of maximum wall deflections induced by braced excavation in clays. Geomech Eng 14(4):315–324

Yang Y, Zhang Q (1997) A hierarchical analysis for rock engineering using artificial neural networks. Rock Mech Rock Eng 30(4):207–222

Yilmaz I (2009) Landslide susceptibility mapping using frequency ratio, logistic regression, artificial neural networks and their comparison: a case study from Kat landslides (Tokat-Turkey). Comput Geosci 35:1125–1138

Young Su K, Byung Tak K (2006) Use of artificial neural networks in the prediction of liquefaction resistance of sands. J Geotech Geoenviron Eng 132(11):1502–1504

Zeng P, Senent S, Jimenez R (2014) Reliability analysis of circular tunnel face stability obeying Hoeke Brown failure criterion considering different distribution types and correlation structures. J Comput Civil Eng 30:04014126

Zhang WG, Goh ATC (2013) Multivariate adaptive regression splines for analysis of geotechnical engineering systems. Comput Geotech 48:82–95

Zhang WG, Goh ATC (2014) Multivariate adaptive regression splines model for reliability assessment of serviceability limit state of twin caverns. Geomech Eng 7(4):431–458

Zhang WG, Goh ATC (2015) Regression models for estimating ultimate and serviceability limit states of underground rock caverns. Eng Geol 188:68–76

Zhang WG, Goh ATC (2016) Evaluating seismic liquefaction potential using multivariate adaptive regression splines and logistic regression. Geomech Eng 10(3):269–284

Zhang WG, Goh ATC, Zhang YM, Chen YM, Xiao Y (2015) Assessment of soil liquefaction based on capacity energy concept and multivariate adaptive regression splines. Eng Geol 188:29–37

Zhang WG, Zhang RH, Goh ATC (2017a) Multivariate adaptive regression splines approach to estimate lateral wall deflection profiles caused by braced excavations in clays. Geotech Geol Eng 36(2):1349–1363

Zhang WG, Zhang YM, Goh ATC (2017b) Multivariate adaptive regression splines for inverse analysis of soil and wall properties in braced excavation. Tunn Undergr Space Technol 64:24–33

Zhang RH, Zhang WG, Goh ATC (2018) Numerical investigation of pile responses caused by adjacent braced excavation in soft clays. Int J Geotech Eng. https://doi.org/10.1080/19386362.2018.1515810

Zhu AX (2000) Mapping soil landscape as spatial continua: the neural network approach. Water Resour Res 36:663–677

Zhu WS, Sui B, Li XJ, Li SC, Wang WT (2008) A methodology for studying the high wall displacement of large scale underground cavern complexes and its applications. Tunn Undergr Space Technol 23:651–664

Chapter 3
Simple MARS Modeling Examples

Three examples consisting of fairly complicated mathematical functions (with single or two variables) are presented to demonstrate the function approximating capacity of MARS. This is followed by an example to evaluate the MARS efficiency in analyzing a hypothetical nonlinear function in which noise (error) is introduced.

3.1 Simple Function Approximation

Figure 3.1 shows a simple example of how MARS would use piecewise linear spline functions to attempt to fit data. The MARS mathematical equation is expressed as

$$y = -44.08 + 4.24 \times BF1 - 3.67 \times BF2 + 6.31 \times BF3 - 2.50 \times BF4 \quad (3.1)$$

Fig. 3.1 Knots and linear splines for a simple MARS example

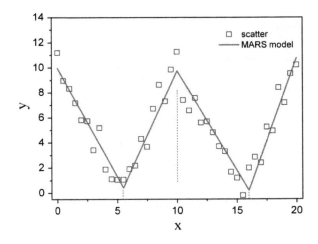

© Science Press and Springer Nature Singapore Pte Ltd. 2019
W. Zhang, *MARS Applications in Geotechnical Engineering Systems*,
https://doi.org/10.1007/978-981-13-7422-7_3

Fig. 3.2 Curve fitting using MARS: **a** a sine function; **b** exponential function

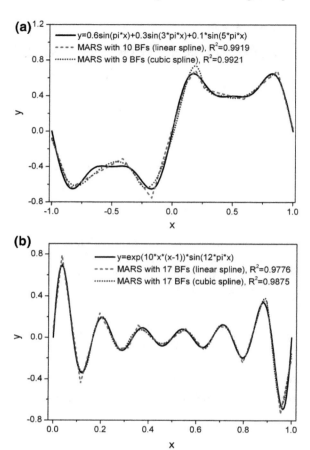

where BF1 = max(0, 16 − x), BF2 = max(0, x − 10), BF3 = max(0, x − 5.5), and BF4 = max(0, 5.5 − x). The knots are located at x = 5.5, 10, and 16. They delimit four intervals where different linear relationships are identified.

In the following example, MARS was used to analyze two complicated nonlinear functions consisting of a single variable:

$$y = 0.6\sin(\pi x) + 0.3\sin(3\pi x) + 0.1\sin(5\pi x) \quad (-1 < x < 1) \tag{3.2}$$

$$y = e^{10x(x-1)}\sin(12\pi x) \quad (0 < x < 1) \tag{3.3}$$

Figure 3.2a and b shows the learning results of the above two functions obtained by MARS. The high coefficient of determination R^2 value indicates that MARS is highly accurate in approximating these two functions.

3.2 Two-Dimensional Approximation

Figure 3.3 shows a two-variable function (Eq. 3.4), which has been widely used for model performance validation.

$$y = \sin(0.83\pi x_1)\cos(1.25\pi x_2) \quad (-1 < x_1, x_2 < 1) \tag{3.4}$$

To approximate this function, two MARS models with 45 BFs of linear and cubic spline functions, respectively, are used as shown in Fig. 3.4a and b. R^2 values of 0.9976 and 0.9991 show that MARS models with sufficient BFs can be used to approximate a two-dimensional function accurately.

Figure 3.5 shows a cowboy hat surface function that has been widely used for validating the performance of neural network and other surrogate models. Both x_1 and x_2 are limited to $[-3, 3]$. A set of data points consisting of 500 training data and 300 testing data was randomly generated using uniform distributions for x_1 and x_2, respectively. The values of z are then calculated from the Equation $z = \sin\left(\sqrt{x_1^2 + x_2^2}\right)$. Chua (2001) found that the evolutionary Bayesian back-propagation (EBBP) predicts well in terms of mean square error (MSE), especially for the testing phase.

To evaluate the accuracy of MARS, the same problem is considered. In the first (forward) phase, a maximum number of 70 BFs of linear spline function with second-order interaction were specified, and subsequently, 28 BFs were pruned from the final

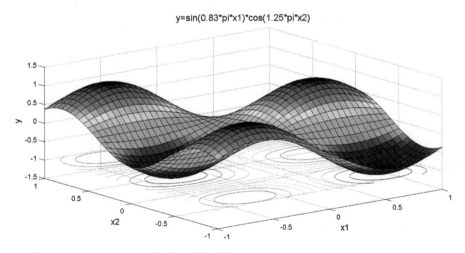

Fig. 3.3 A two-variable function for surface fitting

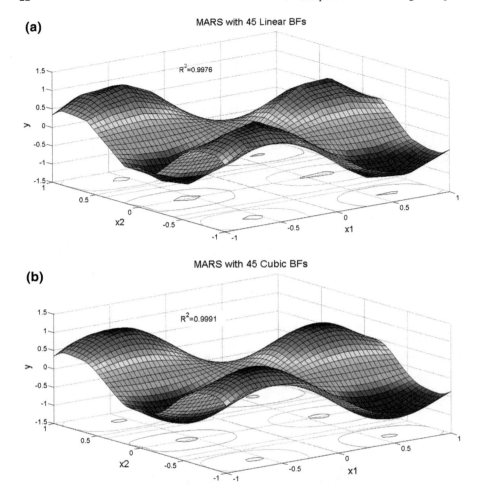

Fig. 3.4 Surface fitting using MARS: **a** using BFs of linear spline; **b** using BFs of cubic spline

MARS model in the second (backward) phase. The summary of the predictions is shown in Table 3.1. Figure 3.6 shows the predictions given by MARS and EBBP. Generally, MARS performs as well as, if not better than the EBBP in terms of MSE especially for the testing phase. In addition, MARS is computationally efficient in terms of processing speed.

Fig. 3.5 Cowboy hat surface

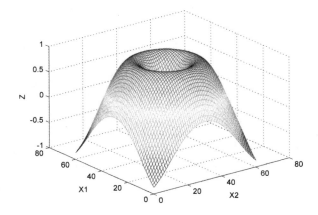

Methods		Training phase	Testing phase
EPPB	MSE ($\times 10^{-3}$)	0.4	0.7
	Coefficient of determination R^2	0.9991	0.9985
MARS	MSE ($\times 10^{-3}$)	0.6	0.8
	Coefficient of determination R^2	0.9974	0.9964

Table 3.1 Comparison of results from EBBP and MARS for fitting cowboy hat

3.3   Function Approximation with Noise

A polynomial function $y = 3x^3$ with Gaussian noise ε is used to verify the generalization capability and accuracy of MARS for the case of a rather large error variance (i.e., noisy data). The x is uniformly distributed between -1 and 1 and ε is normally distributed with mean value of 0. Two cases were considered, one with a smaller error (variance of 0.25) and one with a larger error (variance of 1.0). In addition, two types of spline functions (linear spline functions and cubic spline functions with the maximum number of BFs set as 6) were used. Figure 3.7 shows the scatter plots with the corresponding MARS regression curves of this example. Also displayed in the plot is the exact curve of $y = 3x^3$. The plots show that the MARS approximation almost completely overlays the exact function, yielding a very good fit to the data, even in the case of a rather large error variance.

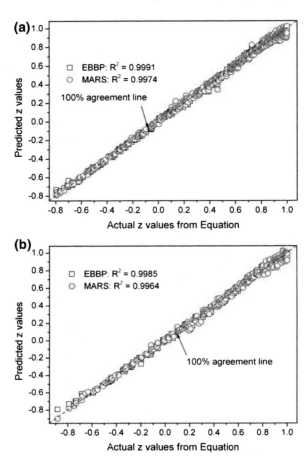

Fig. 3.6 Prediction of cowboy hat function by MARS and EBBP: **a** training data sets, **b** testing data sets

3.4   Summary

This chapter presented three examples consisting of fairly complicated mathematical functions (with single or two variables) to demonstrate the function approximating capacity of MARS, followed by an example to evaluate the MARS efficiency in analyzing a hypothetical nonlinear function in which noise (error) is introduced.

Fig. 3.7 MARS
approximations: **a** large error
variance of $\sigma^2 = 1$; **b** small
error variance of $\sigma^2 = 0.25$

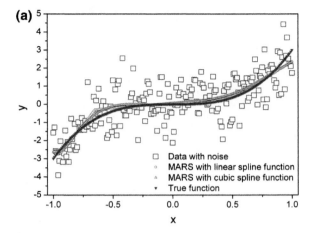

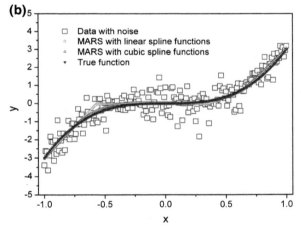

Reference

Chua CG (2001) Prediction of the behavior of braced excavation systems using Bayesian neural networks. Master thesis, Nanyang Technological University, Singapore

Chapter 4
MARS Use in Prediction of Collapse Potential for Compacted Soils

Collapse, defined as the additional deformation of compacted soils when wetted, is believed to be responsible for damage to buildings resting on compacted fills, as well as failure in embankments and earth dams. In this chapter, two surrogate model building techniques, MARS and BPNN, are employed as computational tools to predict the amount of collapse and to investigate the influence of various parameters on the collapse potential. The performances of MARS and BPNN are compared with regard to the predictive accuracy, the computational time, the developed model interpretability. Some useful conclusions are also arrived at.

4.1 Background

All soils settle upon loading. However, unsaturated soils may reduce in volume when inundated under constant applied pressure. The amount of this additional deformation, called "collapse", depends on several factors including the applied pressure, water content, dry density, principal stress ratio, clay content, and compaction method. For consideration of the suction force due to the unsaturated condition, the effective stress principle, used to describe the mechanical behavior of unsaturated soils, fails to predict the collapse phenomena (Jennings and Burland 1962).

Collapse may occur in natural soils as well as in compacted soils and embankments although the mechanism and contributing agents may differ. The influence of different parameters on the amount of collapse has been discussed by many investigators (Lawton et al. 1989; Booth 1975; Barden et al. 1973; Alwail 1990; Dudley 1970). Collapse potential is assessed by different investigators employing different methods. These methods vary from simple empirical equations based on statistical regression (Basma and Tuncer 1992; Nwabuokei and Lovell 1986) to experimental procedures, such as single and double oedometer tests, which have been described by Jennings and Knights (1957), Houston et al. (1988), respectively.

© Science Press and Springer Nature Singapore Pte Ltd. 2019
W. Zhang, *MARS Applications in Geotechnical Engineering Systems*,
https://doi.org/10.1007/978-981-13-7422-7_4

Wetting-induced collapse is believed to be responsible for failure in several earth dams (Leonards and Davidson 1984; Peterson and Iverson 1953), as well as damage to buildings resting on compacted soils, reported by Lawton et al. (2015). Despite the considerable amount of work done in this area, the functional relationship between various soil parameters and the amount of collapse deformation is not well established and the exact interrelationship is still a matter of speculation. The MARS and BPNN methods, as a computational tool, have proven to be capable of establishing a relationship between a series of input data and the corresponding outputs, no matter how complex this implicit relationship may be. Hence, these two methods are employed in this chapter to investigate the collapse potential of unsaturated soils. In order to arrive at robust surrogate models, a comprehensive database is required a priori. Therefore, a more comprehensive database containing 330 sets of data is compiled by Habibagahi and Taherian (2004). Based on the database, the performances of MARS and BPNN are systematically compared, with regard to the predictive accuracy, the computational time (using a PC with 3.0 GHz Intel Core2Quad Q9650 processor, 4 GB RAM), the developed model interpretability.

4.2   The Database

A powerful surrogate model needs a comprehensive database to cover a wide variety of soil types with different initial conditions. Such an estimation model will be capable of predicting collapse potential with good accuracy not only for the training patterns (training data sets) but also for the patterns that the network had not been exposed to during the training process (validation or generalization property). To arrive at this goal, an attempt was made to increase the amount of data in the database by adding data that were available in the literature based on similar test procedures.

A total of 138 sets of data from test results performed on eight different soils and reported by Basma and Tuncer (1992) were added to the database of 192 oedometer tests performed by Habibagahi and Taherian (2004) to arrive at a total of 330 sets of data. The properties of soils as well as the results of oedometric collapse tests reported by Basma and Tuncer (1992) are listed in Tables 4.1 and 4.2, respectively.

A large number of oedometric tests (192 tests) were conducted and the amount of collapse was measured under different initial conditions and at different applied stresses. The properties of soils, details of tested specimens as well as the results of oedometric collapse tests reported by Habibagahi and Taherian (2004) are listed in Tables 4.3, 4.4, and 4.5, respectively.

Table 4.6 lists the ten inputs for BPNN and MARS model to predict the collapse potential.

Table 4.1 Properties of soils tested by Basma and Tuncer (1992)

Soil series	Sand (%)	Silt (%)	Clay (%)	C_U	C_C	LL %	PI %	G_S
S1	40.6	50.5	8.9	17.5	7.2	36.6	12.7	2.74
S2	47.8	47.2	5.0	25.0	1.1	29.1	11.2	2.72
S3	13.3	73.5	13.2	60.0	15.0	57.2	28.9	2.69
S4	19.6	70.4	10.0	11.5	2.9	28.0	7.0	2.77
S5	24.4	49.6	26.0	35.0	0.5	36.0	11.1	2.66
S6	42.1	42.9	15.0	100.0	0.9	28.2	10.6	2.69
S7	84.0	7.0	9.0	6.4	1.6	30.0	3.0	2.63
S8	92.2	5.8	2.0	3.4	1.1	25.0	5.0	2.65

From a total of 330 sets of data, 264 sets were used for training the network. The remaining 66 sets of data were used to test the network (Habibagahi and Taherian 2004). Testing and training data sets are indicated in Tables 4.2 and 4.5. Using the same training and testing data sets, this problem is reanalyzed by means of BPNN and MARS.

4.3 The Developed MARS Models and Modeling Results

The optimal BPNN model is with three hidden neurons while the optimal MARS model adopted 19 BFs of linear spline functions with second-order interaction. A plot of the BPNN and MARS predicted collapse potential values versus the actual values for the training and testing patterns is shown in Fig. 4.1. Comparison between BPNN and MARS in Table 4.7 indicates that the MARS model is slightly more accurate. Table 4.7 also suggests that MARS also outperforms BPNN in processing time.

4.4 Model Interpretability

Table 4.8 lists the BFs and their corresponding equations. 16 of the total 19 BFs are interaction terms with second-order (excluding BF1, BF2 and BF3), indicating that the model is not simply additive and that interaction terms play a significantly important role. The interpretable MARS model is given by

Table 4.2 Results of oedometric collapse tests reported by Basma and Tuncer (1992)

Soil series	Soil specimens	Initial water content (%)	Initial dry density (Mg/m^3)	Applied pressure at wetting (kPa)
A	S1, S2, S3, S4	4.9	1.36, 1.47, 1.56, 1.64	100
A	S5, S6, S7, S8	5.3	1.30, 1.45, 1.50, 1.66	200
A	S9, S10, S11, S12	6	1.32, 1.42, 1.52, 1.61	400
A	S13, S14, S15, S16	5.8	1.30, 1.42, 1.55, 1.61	800
A	S17, S18, S19, S20	9.4	1.23, 1.48, 1.61, 1.73	100
A	S21, S22, S23, S24	9.7	1.28, 1.45, 1.55, 1.72	200
A	S25, S26, S27, S28	9.3	1.24, 1.43, 1.54, 1.68	400
A	S29, S30, S31, S32	9.2	1.24, 1.42, 1.53, 1.69	800
A	S33, S34, S35, S36	12.5	1.32, 1.54, 1.62, 1.84	100
A	S37, S38, S39, S40	11.6	1.28, 1.41, 1.54, 1.70	200
A	S41, S42, S43, S44	12.4	1.25, 1.39, 1.53, 1.69	400
A	S45, S46, S47, S48	12.1	1.26, 1.43, 1.50, 1.68	800
A	S49, S50, S51, S52	15.7	1.40, 1.50, 1.65, 1.84	100
A	S53, S54, S55, S56	15.6	1.25, 1.37, 1.48, 1.55	200
A	S57, S58, S59, S60	15.6	1.28, 1.35, 1.50, 1.56	400
A	S61, S62, S63, S64	16.3	1.25, 1.34, 1.50, 1.58	800
B	S65, S66, S67, S68	5.4	1.36, 1.44, 1.56, 1.65	100
B	S69, S70, S71, S72	6.1	1.34, 1.44, 1.54, 1.62	200
B	S73, S74, S75, S76	5	1.38, 1.49, 1.57, 1.65	400
B	S77, S78, S79, S80	5.5	1.38, 1.48, 1.54, 1.62	800

(continued)

Table 4.2 (continued)

Soil series	Soil specimens	Initial water content (%)	Initial dry density (Mg/m^3)	Applied pressure at wetting (kPa)
B	S81, S82, S83, S84	9.2	1.32, 1.46, 1.62, 1.72	100
B	S85, S86, S87, S88	9.1	1.35, 1.50, 1.60, 1.73	200
B	S89, S90, S91, S92	8.4	1.36, 1.49, 1.65, 1.75	400
B	S93, S94, S95, S96	9.1	1.36, 1.46, 1.57, 1.72	800
B	S97, S98, S99, S100	12.4	1.30, 1.45, 1.55, 1.77	100
B	S101, S102, S103, S104	12.4	1.32, 1.44, 1.60, 1.77	200
B	S105, S106, S107, S108	12.4	1.33, 1.40, 1.51, 1.74	400
B	S109, S110, S111, S112	12.4	1.30, 1.41, 1.58, 1.69	800
B	S113, S114, S115, S116	16.9	1.40, 1.52, 1.59, 1.75	100
B	S117, S118, S119, S120	16.9	1.34, 1.47, 1.60, 1.75	200
B	S121, S122, S123, S124	16.9	1.34, 1.51, 1.62, 1.80	400
B	S125, S126, S127, S128	16.9	1.37, 1.47, 1.61, 1.77	800
C	S129, S130, S131, S132	6	1.45, 1.57, 1.64, 1.77	100
C	S133, S134, S135, S136	6	1.45, 1.57, 1.67, 1.79	200
C	S137, S138, S139, S140	6	1.42, 4.53, 1.67, 1.78	400
C	S141, S142, S143, S144	6	1.48, 1.59, 1.66, 1.80	800
C	S145, S146, S147, S148	9.3	1.43, 1.58, 1.75, 1.91	100
C	S149, S150, S151, S152	9.3	1.40, 1.60, 1.75, 1.89	200

(continued)

Table 4.2 (continued)

Soil series	Soil specimens	Initial water content (%)	Initial dry density (Mg/m^3)	Applied pressure at wetting (kPa)
C	S153, S154, S155, S156	9.3	1.36, 1.52, 1.65, 1.75	400
C	S157, S158, S159, S160	9.3	1.41, 1.51, 1.61, 1.72	800
C	S161, S162, S163, S164	12.2	1.50, 1.57, 1.73, 1.90	100
C	S165, S166, S167, S168	12.2	1.45, 1.58, 1.68, 1.85	200
C	S169, S170, S171, S172	12.2	1.46, 1.57, 1.69, 1.88	400
C	S173, S174, S175, S176	12.2	1.45, 1.54, 1.70, 1.84	800
C	S177, S178, S179, S180	15.7	1.52, 1.65, 1.71, 1.78	100
C	S181, S182, S183, S184	15.7	1.45, 1.63, 1.70, 1.77	200
C	S185, S186, S187, S188	15.7	1.46, 1.55, 1.65, 1.74	400
C	S189, S190, S191, S192	15.7	1.44, 1.58, 1.70, 1.76	800

Table 4.3 Properties of tested soils

Soil series	Sand (%)	Silt (%)	Clay (%)	C_U	C_C	LL %	PI %	G_S
A	13.0	75.0	12.0	16.7	1.4	22.6	5.0	2.68
B	32.0	52.0	16.0	50.0	1.8	24.2	8.0	2.68
C	35.0	52.0	13.0	35.0	2.4	28.2	3.0	2.68

$$
\begin{aligned}
CP(\%) = {} & -0.2524 + 0.7743 \times BF1 + 0.2376 \times BF2 + 0.2131 \times BF3 - 0.1755 \times BF4 \\
& - 0.057 \times BF5 - 0.0184 \times BF6 + 0.01 \times BF7 + 0.0032 \times BF8 + 0.0206 \times BF9 \\
& + 0.0343 \times BF10 - 0.001 \times BF11 + 0.0843 \times BF12 + 0.0645 \times BF13 \\
& - 0.4218 \times BF14 - 0.1665 \times BF15 + 0.0007 \times BF16 + 0.0019 \times BF17 \\
& - 0.0026 \times BF18 - 0.7086 \times BF19
\end{aligned}
$$

$$(4.1)$$

Table 4.4 Details of tested specimens

Sample no.	Initial water content: %	Initial dry unit weight: kN/m³	Pressure at wetting: kPa	Collapse potential: %	NN pattern type
S1	4.9	13.64	100	14.1	Training
S2	4.9	14.72	100	8.1	Training
S3	4.9	15.6	100	4.2	Training
S4	4.9	16.48	100	3.9	Training
S5	5.3	13.15	200	17.3	Testing
S6	5.3	14.62	200	11.6	Training
S7	5.3	15.21	200	8.2	Training
S8	5.3	16.68	200	6.3	Training
S9	6	13.44	400	17.1	Training
S10	6	14.32	400	13.5	Testing
S11	6	15.3	400	9	Training
S12	6	16.28	400	6.1	Training
S13	5.8	13.14	800	15.2	Training
S14	5.8	14.42	800	14.6	Training
S15	5.8	15.6	800	11.3	Testing
S16	5.8	16.19	800	9.3	Training
S17	9.4	12.56	100	17.6	Training
S18	9.4	14.91	100	4.5	Training
S19	9.4	16.28	100	1.7	Training
S20	9.4	17.46	100	0	Testing
S21	9.7	12.85	200	17.6	Training
S22	9.7	14.62	200	8.7	Training
S23	9.7	15.6	200	3.7	Training
S24	9.7	17.27	200	0.4	Training
S25	9.3	12.56	400	11.2	Testing
S26	9.3	14.52	400	11.4	Training
S27	9.3	15.5	400	6.4	Training
S28	9.3	16.97	400	1	Training
S29	9.2	12.56	800	10.9	Training
S30	9.2	14.32	800	13.3	Testing
S31	9.2	15.6	800	7.7	Training
S32	9.2	16.97	800	4.1	Training

(continued)

Table 4.4 (continued)

Sample no.	Initial water content: %	Initial dry unit weight: kN/m^3	Pressure at wetting: kPa	Collapse potential: %	NN pattern type
S33	12.5	13.44	100	13.9	Training
S34	12.5	15.6	100	1.9	Training
S35	12.5	16.38	100	0.2	Testing
S36	12.5	18.54	100	0	Training
S37	11.6	13.05	200	14.1	Training
S38	11.6	14.22	200	11.3	Training
S39	11.6	15.6	200	5.5	Training
S40	11.6	17.17	200	0.3	Testing
S41	12.4	12.65	400	11	Training
S42	12.4	14.03	400	11.2	Training
S43	12.4	15.5	400	4.8	Training
S44	12.4	17.07	400	0.4	Training
S45	12.1	12.85	800	5.6	Testing
S46	12.1	14.52	800	7.4	Training
S47	12.1	15.21	800	6.4	Training
S48	12.1	16.97	800	2.4	Training
S49	15.7	14.03	100	10.4	Training
S50	15.7	15.01	100	5.2	Testing
S51	15.7	16.59	100	0.1	Training
S52	15.7	18.44	100	0	Training
S53	14.6	12.65	200	9	Training
S54	14.6	13.83	200	9.3	Training
S55	14.6	14.91	200	7.4	Testing
S56	14.6	15.6	200	4.7	Training
S57	15.6	12.75	400	4.6	Training
S58	15.6	13.54	400	4.9	Training
S59	15.6	15.11	400	5.1	Training
S60	15.6	15.7	400	4.5	Testing
S61	16.3	12.66	800	0.6	Training
S62	16.3	13.54	800	1.3	Training
S63	16.3	15.01	800	0.3	Training
S64	16.3	15.89	800	0.2	Training

(continued)

Table 4.4 (continued)

Sample no.	Initial water content: %	Initial dry unit weight: kN/m³	Pressure at wetting: kPa	Collapse potential: %	NN pattern type
S65	5.4	13.54	100	10.4	Testing
S66	5.4	14.32	100	8.2	Training
S67	5.4	15.5	100	2.1	Training
S68	5.4	16.38	100	2.9	Training
S69	6.1	13.54	200	13	Training
S70	6.1	14.32	200	10	Testing
S71	6.1	15.31	200	7.1	Training
S72	6.1	16.09	200	5.1	Training
S73	5	13.73	400	14.3	Training
S74	5	14.81	400	11.2	Training
S75	5	15.6	400	8.4	Testing
S76	5	16.38	400	8.3	Training
S77	5.5	13.73	800	13.2	Training
S78	5.5	14.72	800	12.7	Training
S79	5.5	15.3	800	11.1	Training
S80	5.5	16.09	800	12.5	Testing
S81	9.2	13.15	100	15	Training
S82	9.2	14.81	100	5.5	Training
S83	9.2	16.19	100	0.7	Training
S84	9.2	17.17	100	0	Training
S85	9.1	13.54	200	15	Testing
S86	9.1	15.01	200	9	Training
S87	9.1	15.99	200	0.2	Training
S88	9.1	17.27	200	0.4	Training
S89	8.4	13.54	400	14.4	Training
S90	8.4	14.81	400	11.4	Testing
S91	8.4	16.38	400	3.8	Training
S92	8.4	17.27	400	1.2	Training
S93	9.1	13.64	800	11.1	Training
S94	9.1	14.62	800	11.1	Training
S95	9.1	15.7	800	8	Testing
S96	9.1	17.17	800	2.1	Training

(continued)

Table 4.4 (continued)

Sample no.	Initial water content: %	Initial dry unit weight: kN/m^3	Pressure at wetting: kPa	Collapse potential: %	NN pattern type
S97	12.4	13.15	100	14.3	Training
S98	12.4	14.62	100	9.3	Training
S99	12.4	15.6	100	0	Training
S100	12.4	17.76	100	0.1	Testing
S101	12.4	13.34	200	10.5	Training
S102	12.4	14.52	200	5.6	Training
S103	12.4	16.09	200	1.4	Training
S104	12.4	17.76	200	0.1	Training
S105	12.4	13.44	400	10.1	Testing
S106	12.4	14.03	400	9.9	Training
S107	12.4	15.21	400	6.6	Training
S108	12.4	17.46	400	0.2	Training
S109	12.4	13.05	800	8.1	Training
S110	12.4	14.22	800	9.1	Testing
S111	12.4	15.99	800	5.1	Training
S112	12.4	16.87	800	1.4	Training
S113	16.9	14.13	100	10.8	Training
S114	16.9	15.03	100	7	Training
S115	16.9	15.99	100	1.1	Testing
S116	16.9	17.56	100	0	Training
S117	16.9	13.54	200	12.4	Training
S118	16.9	14.81	200	8.9	Training
S119	16.9	16.09	200	3.4	Training
S120	16.9	17.56	200	0	Testing
S121	16.9	13.54	400	5	Training
S122	16.9	15.21	400	5.4	Training
S123	16.9	16.28	400	2.4	Training
S124	16.9	18.05	400	0	Training
S125	16.9	13.83	800	0.1	Testing
S126	16.9	14.72	800	1.3	Training
S127	16.9	16.19	800	0.1	Training
S128	16.9	17.76	800	0	Training

(continued)

Table 4.4 (continued)

Sample no.	Initial water content: %	Initial dry unit weight: kN/m^3	Pressure at wetting: kPa	Collapse potential: %	NN pattern type
S129	6	14.42	100	10.9	Training
S130	6	15.6	100	6.8	Testing
S131	6	16.28	100	2.9	Training
S132	6	17.6	100	0.8	Training
S133	6	14.42	200	13.8	Training
S134	6	15.6	200	8.5	Training
S135	6	16.58	200	4.3	Testing
S136	6	17.76	200	1.7	Training
S137	6	14.13	400	13.6	Training
S138	6	15.21	400	12.2	Training
S139	6	16.58	400	6.2	Training
S140	6	17.66	400	2.6	Testing
S141	6	14.72	800	13	Training
S142	6	15.79	800	12.6	Training
S143	6	16.48	800	7	Training
S144	6	17.85	800	5.4	Training
S145	9.2	14.22	100	4.7	Testing
S146	9.2	15.7	100	1.8	Training
S147	9.2	17.37	100	0	Training
S148	9.2	18.93	100	0	Training
S149	9.2	13.93	200	13.4	Training
S150	9.2	15.99	200	5.4	Testing
S151	9.2	17.37	200	0.5	Training
S152	9.2	18.74	200	0	Training
S153	9.2	13.54	400	7.4	Training
S154	9.2	15.11	400	8.4	Training
S155	9.2	16.38	400	5.6	Testing
S156	9.2	17.36	400	0.9	Training
S157	9.2	14.13	800	4.5	Training
S158	9.2	15.21	800	5.3	Training
S159	9.2	15.99	800	5.5	Training
S160	9.2	17.17	800	3.3	Testing

(continued)

Table 4.4 (continued)

Sample no.	Initial water content: %	Initial dry unit weight: kN/m^3	Pressure at wetting: kPa	Collapse potential: %	NN pattern type
S161	12.2	14.91	100	10.1	Training
S162	12.2	15.7	100	6.4	Training
S163	12.2	17.27	100	0	Training
S164	12.2	18.93	100	0	Training
S165	12.2	14.52	200	5.5	Testing
S166	12.2	15.79	200	5.2	Training
S167	12.2	16.78	200	0	Training
S168	12.2	18.44	200	0	Training
S169	12.2	14.62	400	5	Training
S170	12.2	15.7	400	3.6	Testing
S171	12.2	16.87	400	2	Training
S172	12.2	18.74	400	0	Training
S173	12.2	14.52	800	1.5	Training
S174	12.2	15.4	800	2.3	Training
S175	12.2	16.97	800	2.3	Testing
S176	12.2	18.34	800	0.3	Training
S177	15.7	15.3	100	1.7	Training
S178	15.7	16.58	100	0	Training
S179	15.7	17.17	100	0	Training
S180	15.7	17.85	100	0	Testing
S181	15.7	14.62	200	0.9	Training
S182	15.7	16.38	200	0.3	Training
S183	15.7	17.07	200	0.1	Training
S184	15.7	17.76	200	0	Training
S185	15.7	14.72	400	0.1	Testing
S186	15.7	15.6	400	0.1	Training
S187	15.7	16.58	400	0	Training
S188	15.7	17.46	400	0	Training
S189	15.7	14.52	800	0	Training
S190	15.7	15.89	800	0	Testing
S191	15.7	17.07	800	0	Training
S192	15.7	17.66	800	0	Training

Table 4.5 Oedometric test results

Soil Series	Initial water content: %	Initial dry unit weight: kN/m^3	Pressure at wetting: kPa	Collapse potential: %	NN pattern type
S1	4	15	400	12.5	Training
S1	6	15	400	10.1	Training
S1	8	15	400	12.5	Testing
S1	12	15	400	11.9	Training
S1	16	15	400	9.1	Training
S1	20	15	400	7.5	Training
S1	6	13.1	400	14.4	Training
S1	6	14	400	13.2	Testing
S1	6	15	400	10.2	Training
S1	6	15.9	400	7.8	Training
S1	6	16.8	400	4.2	Training
S1	6	17.8	400	1.3	Training
S1	6	18.7	400	0	Testing
S1	6	15	200	7.1	Training
S1	6	15	400	12.7	Training
S1	6	15	800	15	Training
S1	6	15	1600	15.6	Training
S1	6	15	3200	15.8	Testing
S2	4	15.4	400	14.8	Training
S2	6	15.4	400	13.3	Training
S2	8	15.4	400	11.7	Training
S2	12	15.4	400	8.7	Training
S2	16	15.4	400	5	Testing
S2	20	15.4	400	0.1	Training
S2	6	13.5	400	21.3	Training
S2	6	14.5	400	18.7	Training
S2	6	15.4	400	13.6	Training
S2	6	16.4	400	9.6	Testing
S2	6	17.4	400	6	Training
S2	6	18.3	400	1	Training

(continued)

Table 4.5 (continued)

Soil Series	Initial water content: %	Initial dry unit weight: kN/m^3	Pressure at wetting: kPa	Collapse potential: %	NN pattern type
S2	6	19.3	400	0	Training
S2	6	15.4	200	8.5	Training
S2	6	15.4	400	13.6	Testing
S2	6	15.4	800	14.7	Training
S2	6	15.4	1200	17.5	Training
S2	6	15.4	3600	17.9	Training
S3	4	13.6	400	19.2	Training
S3	6	13.6	400	17.5	Testing
S3	8	13.6	400	16.2	Training
S3	12	13.6	400	15	Training
S3	16	13.6	400	13.2	Training
S3	20	13.6	400	12	Training
S3	6	11.9	400	22.7	Testing
S3	6	12.8	400	20	Training
S3	6	13.6	400	17.5	Training
S3	6	14.5	400	9.5	Training
S3	6	15.3	400	6.3	Training
S3	6	16.2	400	3.3	Testing
S3	6	17	400	0.1	Training
S3	6	13.6	200	12	Training
S3	6	13.6	400	17.5	Training
S3	6	13.6	800	19	Training
S3	6	13.6	1600	21.6	Testing
S3	6	13.6	3200	21.9	Training
S4	4	13.8	400	16.8	Training
S4	8	13.8	400	15.1	Training
S4	12	13.8	400	14.3	Training
S4	16	13.8	400	7	Testing
S4	20	13.8	400	9.7	Training
S4	6	12	400	21.3	Training
S4	6	12.9	400	19.5	Training
S4	6	13.8	400	16.6	Testing

(continued)

Table 4.5 (continued)

Soil Series	Initial water content: %	Initial dry unit weight: kN/m^3	Pressure at wetting: kPa	Collapse potential: %	NN pattern type
S4	6	14.6	400	12	Training
S4	6	15.5	400	7.5	Training
S4	6	16.3	400	5.2	Training
S4	6	17.2	400	3.7	Training
S4	6	13.8	200	12	Testing
S4	6	13.8	400	16.5	Training
S4	6	13.8	800	15.1	Training
S4	6	13.8	1600	20.8	Training
S4	6	13.8	3200	23	Training
S5	4	13	400	22.6	Testing
S5	6	13	400	21.1	Training
S5	8	13	400	19.3	Training
S5	12	13	400	19.2	Training
S5	16	13	400	14.9	Training
S5	20	13	400	11	Testing
S5	6	11.4	400	23.2	Training
S5	6	12.2	400	24.1	Training
S5	6	13	400	22.2	Training
S5	6	13.9	400	16.1	Training
S5	6	14.7	400	15.8	Testing
S5	6	15.5	400	11.9	Training
S5	6	13	200	17	Training
S5	6	13	400	22	Training
S5	6	13	800	21.2	Testing
S5	6	13	1600	23.2	Training
S5	6	13	3200	24.5	Training
S6	4	14.6	400	24.5	Training
S6	6	14.6	400	22.5	Training
S6	8	14.6	400	18.6	Testing
S6	12	14.6	400	16.3	Training
S6	16	14.6	400	16	Training
S6	20	14.6	400	14	Training

(continued)

Table 4.5 (continued)

Soil Series	Initial water content: %	Initial dry unit weight: kN/m^3	Pressure at wetting: kPa	Collapse potential: %	NN pattern type
S6	6	12.8	400	26.4	Training
S6	6	13.7	400	25.1	Testing
S6	6	14.6	400	20.2	Training
S6	6	15.6	400	16.5	Training
S6	6	16.5	400	16.1	Training
S6	6	17.4	400	9.4	Training
S6	6	18.3	400	9	Testing
S6	6	14.6	200	14.9	Training
S6	6	14.6	400	19.9	Training
S6	6	14.6	800	23	Training
S6	6	14.6	1600	25.7	Training
S6	6	14.6	3200	26.4	Testing
S7	6	18.2	200	0	Training
S7	6	18.2	400	0.1	Training
S7	6	18.2	800	0.1	Training
S7	6	18.2	1600	1.5	Training
S7	6	18.2	3200	4.2	Testing
S7	3	18.2	400	1.2	Training
S7	6	18.2	400	0	Training
S7	9	18.2	400	0	Training
S7	12	18.2	400	0	Testing
S7	6	15.7	400	6.6	Training
S7	6	16.3	400	4.3	Training
S7	6	17.7	400	1	Training
S7	6	18.2	400	0	Training
S7	6	19.2	400	0	Testing
S8	6	16.9	200	0	Training
S8	6	16.9	400	0.9	Training
S8	6	16.9	800	2.1	Training
S8	6	16.9	1600	3.1	Training
S8	6	16.9	3200	6.5	Testing
S8	0	16.9	400	2.7	Training

(continued)

Table 4.5 (continued)

Soil Series	Initial water content: %	Initial dry unit weight: kN/m³	Pressure at wetting: kPa	Collapse potential: %	NN pattern type
S8	3	16.9	400	1	Training
S8	6	16.9	400	0.5	Training
S8	9	16.9	400	0	Training
S8	12	16.9	400	0	Testing
S8	6	14.6	400	6	Training
S8	6	15.1	400	5.1	Training
S8	6	16.4	400	2.1	Training
S8	6	16.9	400	1	Training
S8	6	17.8	400	0	Testing

Table 4.6 Summary of collapse potential input variables and output

Inputs and output	Parameters and Parameter descriptions	
Input variables	Sand content: Sand (%)	Variable 1 ($\times 1$)
	Silt content: Silt (%)	Variable 2 ($\times 2$)
	Clay content: Clay (%)	Variable 3 ($\times 3$)
	Coefficient of uniformity: C_U	Variable 4 ($\times 4$)
	Coefficient of curvature: C_C	Variable 5 ($\times 5$)
	Liquid limit: LL	Variable 6 ($\times 6$)
	Plasticity index: PI	Variable 7 ($\times 7$)
	Initial water content: ω (%)	Variable 8 ($\times 8$)
	Initial dry unit weight: γ_d (kN/m³)	Variable 9 ($\times 9$)
	Pressure at wetting: P_w (kPa)	Variable 10 ($\times 10$)
Output	Collapse potential CP (%)	

4.5 Parameter Relative Importance

The ANOVA parameter relative importance assessment indicates that the two most significant variables are P_w (Pressure at wetting) and C_U (Coefficient of uniformity). For brevity, the ANOVA decomposition data have been omitted and this part would be explained in the following chapters for more complicated geotechnical engineering problems.

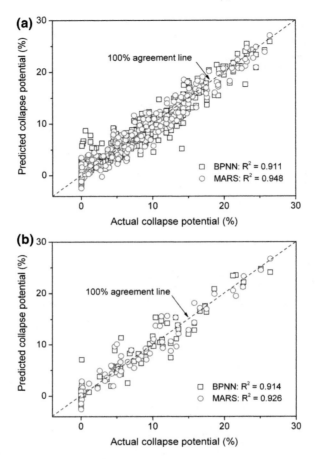

Fig. 4.1 Performance of MARS model for CP: **a** training; **b** testing

Table 4.7 Comparison of performance measures for BPNN and MARS

Methods	Processing time (s)	Accuracy					
		Training sets			Testing sets		
		R^2	MSE	MAE	R^2	MSE	MAE
MARS	154.02	0.948	2.556	1.282	0.926	3.715	1.524
BPNN	2.16	0.911	4.409	1.514	0.914	4.329	1.507

Table 4.8 Basis functions and corresponding equations of MARS model for CP prediction

Basis function	Equation
BF1	$\max(0, 17.07 - \gamma_d)$
BF2	$\max(0, C_U - 50)$
BF3	$\max(0, 50 - C_U)$
BF4	$BF3 \times \max(0, PI - 11.2)$
BF5	$BF3 \times \max(0, 11.2 - PI)$
BF6	$\max(0, P_w - 800)$
BF7	$\max(0, 800 - P_w)$
BF8	$BF7 \times \max(0, 16.09 - \gamma_d)$
BF9	$BF3 \times \max(0, LL - 25)$
BF10	$BF3 \times \max(0, 25 - LL)$
BF11	$BF7 \times \max(0, \omega - 5.4)$
BF12	$\max(0, 16.9 - \omega) \times \max(0, 12.7 - PI)$
BF13	$BF1 \times \max(0, PI - 5)$
BF14	$BF1 \times \max(0, 5 - PI)$
BF15	$\max(0, 16.9 - \omega) \times \max(0, \gamma_d - 13.5)$
BF16	$BF7 \times \max(0, 12.7 - PI)$
BF17	$\max(0, 16.9 - \omega) \times \max(0, P_w - 400)$
BF18	$\max(0, 16.9 - \omega) \times \max(0, 400 - P_w)$
BF19	$\max(0, \omega - 16.9) \times \max(0, 8.9 - Clay)$

4.6 Summary

This chapter examined the MARS and BPNN approaches as surrogate modeling methods in predicting the amount of collapse and investigating the influence of various parameters on the collapse potential. The performances of MARS and BPNN are comprehensively compared with regard to the predictive accuracy, the computational time, the developed model interpretability. Some useful conclusions for the use of the two methods are also arrived at.

References

Alwail T (1990) Mechanism and effect of fines on the collapse of compacted sandy soils. Ph.D. thesis, Washington State University, Pullman, Washington
Barden L, Mcgown A, Collins K (1973) The collapse mechanism in partly saturated soil. Eng Geol 7(1):49–60
Basma AA, Tuncer ER (1992) Evaluation and control of collapsible soils. J Geotech Eng 118(10):1491–1504

Booth AR (1975) The factors influencing collapse settlement in compacted soils. In: Proceedings of sixth regional conference for Africa on sol mechanics and foundation engineering, vol 2, South African Institute of Civil Engineers, pp 57–63

Dudley JH (1970) Review of collapsing soils. J Soil Mech Found Div 97(3):925–947

Habibagahi G, Taherian M (2004) Prediction of collapse potential for compacted soils using artificial neural networks. Scientia Iranica 11(1–2):1–20

Houston SL, Houston WN, Spadola DJ (1988) Prediction of field collapse of soils due to wetting. J Geotech Eng 114(1):40–58

Jennings JE, Burland JB (1962) Limitations to the use of effective stresses in partly saturated soils. Geotechnique 12(2):125–144

Jennings JE, Knight K (1957) The prediction of total heave from the double oedometer test. Trans Symp Eng 13–19

Lawton EC, Fragaszy RJ, Hardcastle JH (1989) Collapse of compacted clayey sand. J Geotech Eng 115(9):1252–1267

Lawton EC, Fragaszy RJ, Hetherington MD (2015) Review of wetting-induced collapse in compacted soil. J Geotech Eng 118(9):1376–1394

Leonards GA, Davidson LW (1984) Reconsideration of failure initiating mechanisms for Teton Dam

Nwabuokei SO, Lovell CW (1986) Compressibility and settlement of compacted fills. Astm Special Technical Publication, pp 184–202

Peterson R, Iverson NL (1953) Study of several low earth dam failures. In: Proceedings of Third International Conference on Soil Mechanics and Foundation Engineering, vol 2, pp 273–276

Chapter 5
MARS Use in Prediction of Diaphragm Wall Deflections in Soft Clays

For excavations in built-up areas with deep deposits of soft clays, it is essential to control ground movements to minimize damage to adjacent structures and facilities. This is commonly carried out by controlling the deflections of the retaining wall system. The limiting wall deflection or serviceability limit state is typically taken to be a percentage of the excavation height. In this study, extensive plane strain finite element analyses have been carried out to examine the excavation-induced wall deflections for a deep deposit of soft clay supported by diaphragm walls and bracing. Based on the numerical results, the BPNN and MARS approaches were used to develop the equations for estimating the maximum wall deflection.

5.1 Background

One of the key concerns in constructing an underground facility within a built-up environment is the impact of the associated ground movements on adjacent buildings. For excavations in the ground that comprises of thick soft clays overlying stiff clay, braced walls are usually used to minimize the ground movements. It is common to extend the wall length into the stiff clay layer to prevent basal heave failure and to reduce the movement of the wall toe. To ensure the serviceability limit state is satisfied, a common design criterion is to limit the maximum wall deflection to a fraction of the excavation depth H_e, typically in the range of 0.5–1.5% of H_e. Unnecessarily, severe restrictions may lead to uneconomic design. Therefore, reliable estimates of wall deflections under working conditions are essential.

The finite element method and the empirical/semiempirical method are two common approaches for estimating wall deflections induced by excavation. The finite element method is widely employed to model complex soil–structure interaction problems. For excavations in soft clays, the Mohr–Coulomb constitutive relationship is commonly used to model the clay stress–strain behavior, with no consideration of the soil small strain effect. The importance of modeling the soil small strain behavior for many geotechnical problems has been highlighted by Burland (1989)

© Science Press and Springer Nature Singapore Pte Ltd. 2019
W. Zhang, *MARS Applications in Geotechnical Engineering Systems*,
https://doi.org/10.1007/978-981-13-7422-7_5

and Jardine et al. (1986). The influence of the soil small strain effect on excavation problems which has been investigated through finite element analysis with some advanced small strain constitutive models (Benz 2007; Osman and Bolton 2006; Kung et al. 2009) showed improvements in the predictions of wall deflection and ground movement.

Empirical and semiempirical methods involve interpolating from a published empirical database or numerical analyses using finite elements. Several empirical and semiempirical methods are available for estimating the excavation-induced maximum wall deflection (Mana and Clough 1981; Wong and Broms 1989; Clough and O'Rourke 1990; Hashash and Whittle 1996; Addenbrooke et al. 2000; Kung et al. 2007a, b; Goh et al. 2013, 2017a, b; Zhang et al. 2015a, b; Wong et al. 1997; Moormann 2004; Wang et al. 2005; Bryson and Zapata-Medina 2012; Hsieh and Ou 2016). However, many empirical methods that have been proposed for estimating wall movements assume that the wall is "floating" in the soft clay, without restraint at the wall toe. This study focuses on the specific situation of the braced wall penetrating into the stiff stratum, since as mentioned previously it is common to extend the wall length into the stiff clay layer to prevent basal heave failure and to reduce the movement of the wall toe.

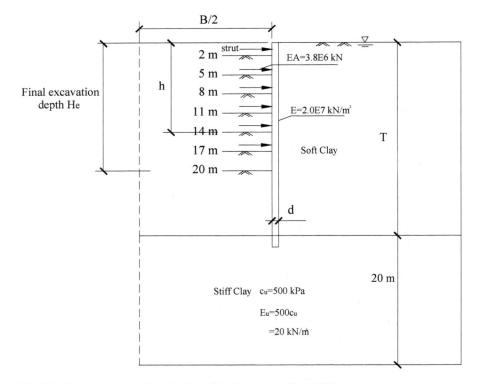

Fig. 5.1 Cross-sectional soil and wall profile (figure from Xuan 2009)

Table 5.1 Summary of a range of finite element parameters for braced excavation analysis

Parameter	Parameter description	Range
c_u/σ_v'	Soil shear strength ratio	0.21, 0.25, 0.29, 0.34
E_{50}/c_u	Soil stiffness	100, 200, 300
γ (kN/m^3)	Soil unit weight	15, 17, 19
T (m)	Soft clay thickness	25, 30, 35
B (m)	Excavation width	20, 30, 40, 50, 60
h (m)	Excavation depth	8, 11, 14, 17, 20
EI ($\times$ 10^6 kN m^2/m)	Wall stiffness	0.36, 1.21, 2.88, 5.63

In this chapter, the database used for surrogate model development is based on Xuan (2009). He carried out extensive plane strain finite element analyses to examine the excavation-induced wall deflections for a deep deposit of soft clay supported by diaphragm walls and bracing. The cross-sectional soil and wall profile are shown in Fig. 5.1. The major parameters influencing excavation performance and the ranges of these parameters are shown in Table 5.1. Because of symmetry, only half of the cross section was considered.

The following sections describe the use of MARS model for relating the maximum wall deflection to various parameters such as the excavation geometry, soil strength and stiffness parameters and the wall stiffness.

5.2 The Database

Finite element analyses were carried out for a total of 1120 cases to determine the maximum diaphragm wall deflection. Of the 1120 cases, 840 patterns were randomly chosen as the training data sets and the remaining as the testing sets for the BPNN and MARS analyses, as listed in Tables 5.2 and 5.3, respectively.

5.3 The Developed MARS Models and Modeling Results

The optimal BPNN structure consisted of seven input neurons, three hidden neurons, and the output neuron representing the maximum wall deflection. The optimal MARS model consisted of 22 BFs of linear spline functions with second-order interaction. A plot of the MARS and BPNN predicted wall deflection values versus the FEM calculated values for the training and testing patterns is shown in Fig. 5.2. The results in Table 5.4 indicate that the BPNN gives slightly better predictions than MARS. However, Table 5.4 also suggests that MARS outperforms BPNN in computational speed.

Table 5.2 A summary of diaphragm wall deflection training data set

B (m)	T (m)	H_e (m)	$\frac{\sigma_t}{\sigma_v}$	$\frac{E_{50}}{c_u}$	$\ln\left(\frac{EI}{\gamma_w h^4_{avg}}\right)$	γ (kN/m³)	δ_{hm} (mm)	B (m)	T (m)	H_e (m)	$\frac{\sigma_t}{\sigma_v}$	$\frac{E_{50}}{c_u}$	$\ln\left(\frac{EI}{\gamma_w h^4_{avg}}\right)$	γ (kN/m³)	δ_{hm} (mm)
40	35	20	0.25	200	6.097	15	566	40	30	14	0.25	200	6.097	15	320
40	35	20	0.29	200	6.097	15	440	40	30	20	0.29	200	6.097	15	316
30	30	20	0.25	100	6.097	15	431	30	35	20	0.29	100	6.097	19	311
40	35	20	0.25	100	6.097	19	406	40	30	20	0.29	100	6.097	17	308
40	30	20	0.29	100	6.097	15	390	40	30	11	0.25	100	6.097	15	307
40	35	17	0.29	200	6.097	15	384	40	30	17	0.25	300	6.097	15	306
40	35	20	0.25	100	7.313	15	379	40	30	14	0.29	100	6.097	15	306
40	30	20	0.25	100	6.097	17	376	40	30	17	0.34	100	6.097	15	305
40	35	20	0.21	200	7.313	15	376	50	35	14	0.29	200	6.097	15	302
40	30	17	0.29	100	6.097	15	371	40	35	17	0.25	100	7.313	17	302
30	30	20	0.29	100	6.097	15	363	30	30	17	0.25	100	6.097	19	300
40	30	17	0.25	200	6.097	15	362	30	30	14	0.25	200	6.097	15	298
40	30	17	0.25	100	6.097	17	354	30	30	20	0.34	100	6.097	15	298
40	35	17	0.25	100	6.097	19	354	40	30	17	0.29	200	6.097	15	296
30	30	20	0.25	100	6.097	17	346	40	30	20	0.25	200	6.097	17	293
40	35	17	0.25	100	7.313	15	345	30	30	20	0.29	200	6.097	15	292
30	30	17	0.29	100	6.097	15	344	40	35	20	0.25	200	6.097	19	291
30	30	17	0.25	200	6.097	15	338	40	35	20	0.29	200	6.097	17	287
40	30	14	0.29	100	6.097	15	332	30	30	20	0.25	200	6.097	17	285
30	30	20	0.25	300	6.097	15	326	40	35	14	0.25	100	6.097	19	284

(continued)

Table 5.2 (continued)

B (m)	T (m)	H_e (m)	$\dfrac{q_t}{\sigma_v}$	$\dfrac{E_{50}}{c_u}$	$\ln\left(\dfrac{EI}{\gamma_w h^4_{\text{avg}}}\right)$	γ (kN/m³)	δ_{hm} (mm)
30	30	17	0.25	100	6.097	17	325
40	30	20	0.25	100	6.097	19	323
30	30	17	0.34	100	6.097	15	279
40	35	17	0.25	200	7.313	15	273
30	30	17	0.25	100	6.097	19	272
40	35	20	0.25	300	7.313	15	272
40	30	17	0.25	200	6.097	17	271
40	35	20	0.25	100	7.313	19	269
40	30	14	0.34	100	6.097	15	268
30	30	14	0.25	300	6.097	15	265
40	35	20	0.34	100	7.313	15	265
40	30	20	0.25	100	7.313	15	262
30	30	20	0.29	300	6.097	15	261
40	30	20	0.29	100	6.097	19	261
50	35	17	0.29	200	6.097	17	261
40	30	14	0.25	100	6.097	19	260
50	35	14	0.25	100	7.313	17	260
40	30	14	0.29	200	6.097	15	257
50	30	20	0.25	200	6.097	19	256
50	35	20	0.25	200	7.313	17	255
40	30	20	0.34	200	6.097	15	252
40	30	11	0.25	100	6.097	17	251
40	30	20	0.29	300	6.097	15	281
50	30	20	0.29	100	6.097	19	281
40	30	20	0.34	100	6.097	17	249
40	35	17	0.25	200	6.097	19	245
30	30	14	0.34	100	6.097	15	244
40	35	17	0.25	300	7.313	15	243
40	35	17	0.29	200	6.097	17	243
30	30	17	0.29	300	6.097	15	241
40	35	17	0.25	100	6.097	19	240
30	30	20	0.25	300	7.313	19	240
40	35	20	0.25	200	6.097	19	239
30	30	20	0.25	300	6.097	17	237
30	30	20	0.29	100	6.097	19	236
40	35	20	0.25	200	7.313	17	236
40	30	17	0.25	300	6.097	17	234
40	35	20	0.25	100	6.097	19	234
30	30	11	0.34	200	6.097	15	233
40	30	20	0.29	200	6.097	17	233
40	30	14	0.25	100	6.097	19	233
50	35	14	0.25	100	7.313	15	232
40	35	14	0.25	200	7.313	15	231
40	30	14	0.25	200	6.097	17	230

(continued)

Table 5.2 (continued)

B (m)	T (m)	H_e (m)	$\dfrac{c_u}{\sigma_v}$	$\dfrac{E_{50}}{c_u}$	$\ln\left(\dfrac{EI}{\gamma_w h^4_{avg}}\right)$	γ (kN/m³)	δ_{hm} (mm)
40	35	20	0.29	200	7.313	15	251
30	30	17	0.25	200	6.097	17	250
40	30	20	0.25	100	7.313	17	229
30	30	17	0.25	100	7.313	15	228
30	30	14	0.29	100	6.097	17	226
50	30	20	0.34	100	6.097	19	226
30	30	20	0.34	100	6.097	17	225
40	30	14	0.29	300	6.097	15	221
40	30	20	0.34	300	6.097	15	221
60	35	20	0.29	200	7.313	17	220
40	30	17	0.29	100	7.313	15	219
30	30	20	0.25	200	6.097	19	219
40	30	17	0.25	100	7.313	17	218
40	30	17	0.25	200	6.097	19	217
30	30	17	0.25	300	6.097	17	216
40	35	14	0.29	100	6.097	19	216
30	30	17	0.34	200	6.097	15	214
40	30	17	0.29	200	6.097	17	213
30	30	20	0.29	200	6.097	17	212
30	30	14	0.25	200	6.097	17	211
40	35	11	0.25	100	6.097	19	211
40	35	20	0.29	200	6.097	19	211

B (m)	T (m)	H_e (m)	$\dfrac{c_u}{\sigma_v}$	$\dfrac{E_{50}}{c_u}$	$\ln\left(\dfrac{EI}{\gamma_w h^4_{avg}}\right)$	γ (kN/m³)	δ_{hm} (mm)
40	30	17	0.34	100	6.097	17	230
40	30	20	0.29	100	7.313	15	229
40	35	11	0.29	200	6.097	15	210
50	35	11	0.25	100	7.313	17	209
40	35	17	0.25	200	7.313	17	208
40	30	20	0.34	100	6.097	19	207
30	30	20	0.25	100	7.313	17	206
30	30	17	0.34	100	6.097	17	206
50	35	20	0.29	200	7.313	17	206
40	30	11	0.25	100	6.097	19	205
40	30	14	0.29	100	6.097	19	205
40	35	20	0.25	300	7.313	17	205
60	30	20	0.25	200	7.313	17	204
40	35	20	0.34	200	7.313	15	204
50	35	14	0.29	200	6.097	17	204
30	30	20	0.34	300	6.097	15	203
40	30	20	0.25	300	6.097	19	202
40	35	14	0.25	100	7.313	19	202
40	30	17	0.34	300	6.097	15	201
40	30	14	0.25	300	7.313	15	201
40	35	17	0.34	100	6.097	19	201
50	30	20	0.29	200	6.097	19	200

(continued)

Table 5.2 (continued)

B (m)	T (m)	H_e (m)	$\frac{q}{\sigma_v}$	$\frac{E_{50}}{c_u}$	$\ln\left(\frac{EI}{\gamma_w h_{avg}^4}\right)$	γ (kN/m³)	δ_{hm} (mm)
40	30	20	0.25	100	7.313	19	210
40	35	20	0.25	200	8.176	15	210
40	30	14	0.25	100	7.313	17	198
40	30	14	0.34	100	6.097	17	198
30	30	17	0.25	200	6.097	19	198
40	30	14	0.34	200	6.097	15	197
40	30	20	0.29	100	7.313	17	197
40	30	20	0.29	300	6.097	17	197
50	30	20	0.29	100	7.313	19	197
40	35	20	0.25	200	7.313	19	195
30	30	17	0.25	100	7.313	17	195
40	35	17	0.29	100	7.313	19	194
40	30	20	0.21	200	7.313	19	193
40	25	20	0.29	100	6.097	19	193
30	30	17	0.29	100	6.097	17	193
60	35	17	0.29	200	7.313	17	191
40	35	20	0.29	200	7.313	17	190
40	30	20	0.25	300	7.313	15	190
60	30	17	0.25	200	7.313	17	190
40	35	17	0.25	200	8.176	15	190
40	30	17	0.34	100	7.313	15	188
40	30	17	0.34	100	6.097	19	188
40	30	11	0.29	100	6.097	17	199
40	35	17	0.25	300	6.097	19	199
40	35	20	0.21	200	8.846	15	187
40	30	17	0.29	100	7.313	17	186
40	35	17	0.34	100	7.313	17	186
50	35	14	0.25	200	7.313	17	186
30	30	17	0.34	300	6.097	15	184
30	30	20	0.34	100	6.097	19	184
40	35	17	0.25	100	8.176	19	184
30	30	17	0.25	200	7.313	15	183
30	30	14	0.29	100	6.097	19	182
40	30	20	0.25	200	7.313	15	181
30	30	17	0.29	300	6.097	17	181
40	30	14	0.29	300	6.097	19	181
30	30	17	0.34	100	7.313	15	180
30	30	14	0.34	200	6.097	15	180
40	30	17	0.21	200	7.313	19	180
40	30	14	0.25	200	6.097	19	180
50	35	17	0.29	200	7.313	17	180
40	30	17	0.25	300	7.313	15	179
40	30	20	0.34	200	6.097	17	179
40	30	20	0.29	100	7.313	19	179

(continued)

Table 5.2 (continued)

B (m)	T (m)	H_e (m)	$\dfrac{\sigma_h}{\sigma_v}$	$\dfrac{E_{50}}{c_u}$	$\ln\left(\dfrac{EI}{\gamma_w h_{avg}^4}\right)$	γ (kN/m³)	δ_{hm} (mm)
40	35	14	0.29	200	6.097	17	188
40	30	20	0.29	200	6.097	19	187
40	30	14	0.29	200	6.097	17	178
40	35	20	0.29	200	8.176	15	178
40	35	17	0.29	200	6.097	19	178
40	35	20	0.34	100	7.313	19	178
40	30	17	0.29	300	6.097	17	177
40	30	14	0.25	100	7.313	19	177
50	30	20	0.25	100	8.176	15	177
50	25	20	0.29	200	6.097	17	176
30	30	14	0.25	100	7.313	17	176
30	30	20	0.29	100	7.313	17	176
40	25	14	0.29	100	6.097	19	175
30	30	14	0.34	100	6.097	17	175
50	35	14	0.34	100	7.313	17	175
30	30	11	0.25	300	6.097	15	174
40	30	20	0.25	200	7.313	17	174
60	30	20	0.29	200	6.097	17	174
40	35	17	0.21	200	8.846	15	173
40	35	20	0.29	100	8.176	19	173
40	25	20	0.25	200	6.097	19	172
50	25	20	0.25	100	7.313	17	171
30	30	14	0.25	300	6.097	17	178
40	30	20	0.25	200	7.313	17	178
40	35	14	0.25	200	7.313	17	171
40	35	17	0.25	200	7.313	19	171
40	30	20	0.25	100	8.176	17	170
50	25	17	0.25	100	7.313	17	169
40	30	11	0.34	100	7.313	17	169
40	30	20	0.34	100	7.313	17	169
30	30	17	0.25	100	8.176	15	168
40	30	14	0.29	100	7.313	15	168
50	30	14	0.34	100	7.313	17	168
30	30	20	0.25	200	6.097	19	168
40	25	17	0.29	200	6.097	19	167
30	30	20	0.29	100	7.313	19	167
40	30	17	0.29	200	6.097	19	167
40	30	17	0.25	100	8.176	15	167
30	30	20	0.34	300	6.097	15	166
40	30	14	0.29	100	7.313	17	166
40	35	17	0.25	200	7.313	17	166
40	35	20	0.25	300	7.313	19	166
30	30	14	0.25	200	7.313	15	165

(continued)

Table 5.2 (continued)

B (m)	T (m)	H_e (m)	$\frac{\sigma_c}{\sigma_v}$	$\frac{E_{50}}{c_u}$	$\ln\left(\frac{EI}{\gamma_w h_{avg}^4}\right)$	γ (kN/m³)	δ_{hm} (mm)
50	25	17	0.29	200	6.097	17	171
40	30	14	0.34	100	7.313	15	171
30	30	17	0.25	300	7.313	15	164
30	30	17	0.25	300	6.097	19	164
30	30	14	0.25	200	6.097	19	163
30	35	20	0.34	200	7.313	17	163
50	35	14	0.29	100	7.313	19	163
40	30	17	0.29	300	6.097	17	162
30	30	20	0.34	200	6.097	17	162
30	30	17	0.25	100	8.176	17	162
40	25	20	0.34	100	6.097	19	161
40	30	20	0.29	300	7.313	15	161
30	30	20	0.25	200	7.313	17	161
40	30	20	0.25	100	8.176	19	161
40	30	14	0.34	100	6.097	19	160
30	30	17	0.25	100	8.176	15	160
30	30	14	0.29	200	6.097	17	160
40	30	17	0.34	200	6.097	17	160
40	35	17	0.29	200	8.176	15	160
40	30	14	0.25	300	7.313	15	159
40	30	17	0.34	100	7.313	17	159
40	35	11	0.25	100	7.313	19	159

B (m)	T (m)	H_e (m)	$\frac{\sigma_c}{\sigma_v}$	$\frac{E_{50}}{c_u}$	$\ln\left(\frac{EI}{\gamma_w h_{avg}^4}\right)$	γ (kN/m³)	δ_{hm} (mm)
30	30	20	0.29	200	7.313	15	165
40	35	14	0.34	100	6.097	19	165
40	25	17	0.34	100	6.097	19	157
40	30	14	0.21	200	7.313	19	157
50	30	20	0.29	100	8.176	19	157
40	35	14	0.34	100	7.313	17	157
40	35	17	0.34	100	7.313	19	157
50	25	14	0.29	200	6.097	17	156
40	30	20	0.25	300	7.313	17	156
40	35	17	0.29	100	8.176	19	156
30	30	17	0.29	200	7.313	15	155
40	30	14	0.25	100	7.313	19	155
40	30	14	0.29	200	7.313	15	154
40	30	20	0.34	200	7.313	15	154
50	25	20	0.29	100	7.313	17	153
50	30	20	0.34	200	6.097	19	153
40	35	14	0.25	300	6.097	19	153
40	25	14	0.25	200	6.097	19	152
40	30	17	0.25	100	8.176	19	152
40	30	20	0.29	300	6.097	19	152
40	30	20	0.34	100	7.313	19	152
40	35	17	0.34	300	7.313	15	152

(continued)

Table 5.2 (continued)

B (m)	T (m)	H_e (m)	$\frac{\sigma_h}{\sigma_v}$	$\frac{E_{50}}{c_u}$	$\ln\left(\frac{EI}{\gamma_w h^4_{avg}}\right)$	γ (kN/m³)	δ_{hm} (mm)	B (m)	T (m)	H_e (m)	$\frac{\sigma_h}{\sigma_v}$	$\frac{E_{50}}{c_u}$	$\ln\left(\frac{EI}{\gamma_w h^4_{avg}}\right)$	γ (kN/m³)	δ_{hm} (mm)
40	30	20	0.25	200	7.313	19	158	60	35	17	0.34	200	7.313	17	152
60	35	14	0.29	200	7.313	17	158	40	35	20	0.34	200	6.097	19	152
30	30	14	0.34	300	6.097	15	151	50	30	20	0.25	300	7.313	19	147
40	30	20	0.29	200	7.313	17	151	50	35	11	0.29	200	6.097	17	147
40	35	20	0.25	200	8.176	19	151	30	25	17	0.29	200	6.097	17	146
50	25	11	0.25	100	7.313	17	150	40	30	14	0.25	200	7.313	17	146
40	25	11	0.29	100	6.097	19	150	40	35	14	0.34	200	7.313	15	146
40	30	17	0.29	300	7.313	15	150	40	35	20	0.34	100	8.176	19	146
30	30	20	0.25	100	8.176	17	150	30	30	14	0.25	300	7.313	15	145
40	30	14	0.25	100	8.176	17	150	30	30	17	0.34	200	6.097	17	145
40	30	20	0.29	100	8.176	17	150	40	30	17	0.25	200	7.313	19	145
40	35	20	0.34	200	7.313	17	150	50	25	14	0.29	100	7.313	17	144
30	30	14	0.25	100	8.176	15	149	40	25	14	0.34	100	6.097	19	144
40	30	14	0.29	100	8.176	15	149	40	30	17	0.25	300	7.313	17	144
40	30	20	0.34	300	6.097	17	149	40	35	20	0.29	200	8.176	17	144
30	30	17	0.29	200	6.097	19	149	30	30	17	0.34	200	7.313	17	143
40	30	11	0.25	100	7.313	19	149	40	30	20	0.21	200	8.176	15	143
40	35	11	0.25	300	7.313	15	149	50	30	20	0.29	200	7.313	19	143
40	25	20	0.25	300	6.097	19	148	40	25	17	0.25	300	6.097	19	142
30	30	20	0.29	100	8.176	15	148	30	30	17	0.34	100	8.176	15	142
40	30	20	0.34	100	8.176	15	148	40	30	17	0.25	100	8.176	17	142
30	30	14	0.29	100	7.313	17	148	30	30	11	0.25	300	6.097	17	142

(continued)

Table 5.2 (continued)

B (m)	T (m)	H_e (m)	$\frac{\sigma_h}{\sigma_v}$	$\frac{E_{50}}{c_u}$	$\ln\left(\frac{EI}{\gamma_w h^4_{avg}}\right)$	γ (kN/m³)	δ_{hm} (mm)
30	30	20	0.29	300	7.313	15	147
40	30	14	0.25	300	6.097	19	147
30	30	20	0.25	200	7.313	19	142
50	35	11	0.34	100	7.313	17	142
40	25	20	0.29	200	6.097	19	141
30	30	20	0.25	300	7.313	17	141
40	30	20	0.29	100	8.176	19	141
40	30	17	0.34	100	7.313	19	141
50	35	17	0.34	200	7.313	17	141
40	35	17	0.29	300	6.097	19	140
30	25	20	0.25	100	7.313	17	140
40	30	17	0.25	200	8.176	15	140
60	30	14	0.29	200	7.313	17	140
30	30	20	0.25	100	8.176	19	140
30	30	14	0.34	100	6.097	19	140
50	35	11	0.25	200	7.313	17	140
40	35	14	0.29	200	6.097	19	139
30	30	20	0.34	200	7.313	15	139
50	35	17	0.29	200	8.176	17	138
30	25	17	0.25	100	7.313	17	138
40	30	14	0.25	100	8.176	19	138
40	35	11	0.25	200	6.097	19	138
40	30	14	0.34	100	7.313	17	142
60	30	20	0.34	200	7.313	17	142
60	30	11	0.25	200	7.313	17	137
30	30	20	0.29	300	6.097	19	137
50	30	20	0.34	100	8.176	19	137
40	25	17	0.29	200	6.097	19	136
30	30	11	0.34	100	6.097	17	136
40	30	20	0.21	300	6.097	17	136
40	35	14	0.29	200	8.176	15	136
40	30	20	0.34	300	8.176	15	135
30	30	20	0.29	200	7.313	17	135
40	30	20	0.25	300	7.313	19	135
40	35	11	0.29	200	6.097	17	135
40	35	20	0.34	300	7.313	15	135
30	30	20	0.25	200	8.176	17	134
50	30	20	0.34	300	6.097	17	134
40	30	11	0.29	200	6.097	17	134
30	30	17	0.29	300	6.097	19	134
40	35	14	0.29	200	7.313	17	134
30	25	14	0.25	100	7.313	17	133
50	30	17	0.34	100	7.313	17	133
40	25	20	0.29	100	7.313	19	133

(continued)

Table 5.2 (continued)

B (m)	T (m)	H_e (m)	$\frac{\sigma}{\sigma_v}$	$\frac{E_{50}}{c_u}$	$\ln\left(\frac{EI}{\gamma_w h_{avg}^4}\right)$	γ (kN/m³)	δ_{hm} (mm)
40	35	14	0.25	200	7.313	19	138
30	30	14	0.29	200	7.313	15	137
40	30	17	0.21	200	8.176	19	133
40	30	17	0.29	100	8.176	19	133
50	25	20	0.25	200	7.313	17	132
50	25	11	0.29	100	7.313	17	132
30	30	11	0.25	100	8.176	15	132
40	30	11	0.29	100	8.176	15	132
40	30	14	0.29	300	7.313	15	132
30	30	20	0.29	100	8.176	17	132
30	30	17	0.25	100	8.176	19	132
30	30	20	0.34	100	7.313	19	132
50	30	20	0.25	200	8.176	19	132
40	35	14	0.34	100	7.313	19	132
30	30	14	0.29	100	8.176	15	131
40	30	20	0.29	200	8.176	15	131
30	30	14	0.25	100	8.176	17	131
30	30	14	0.25	200	7.313	17	131
30	30	14	0.29	300	6.097	17	131
40	30	14	0.29	100	8.176	17	131
40	30	17	0.34	300	6.097	17	131
40	30	20	0.29	200	7.313	19	131
40	30	20	0.34	100	8.176	17	133
30	30	14	0.25	300	6.097	19	133
50	25	11	0.29	200	6.097	17	130
30	30	20	0.34	100	8.176	15	130
30	30	17	0.25	300	7.313	17	130
40	30	14	0.21	200	8.176	17	130
30	30	17	0.25	200	7.313	19	130
50	25	17	0.25	200	7.313	17	129
30	30	17	0.34	200	7.313	15	129
40	30	14	0.29	100	7.313	19	129
40	30	11	0.34	100	6.097	19	129
40	30	17	0.34	200	6.097	19	129
40	35	14	0.25	200	8.176	15	128
30	30	11	0.25	100	7.313	19	128
40	35	11	0.25	200	7.313	17	128
30	35	20	0.25	300	8.176	19	128
40	25	14	0.34	100	7.313	17	127
30	25	14	0.25	300	6.097	19	127
30	30	17	0.21	300	8.176	17	127
40	35	17	0.29	200	8.176	17	127
40	35	20	0.29	300	7.313	19	127
40	30	14	0.34	200	7.313	15	126

(continued)

Table 5.2 (continued)

B (m)	T (m)	H_e (m)	$\dfrac{c_u}{\sigma'_v}$	$\dfrac{E_{50}}{c_u}$	$\ln\left(\dfrac{EI}{\gamma_w h_{avg}^4}\right)$	γ (kN/m³)	δ_{hm} (mm)
40	35	11	0.29	100	7.313	19	131
40	35	17	0.34	100	8.176	19	131
40	35	11	0.21	200	8.846	15	126
40	25	11	0.25	200	6.097	19	125
40	30	17	0.25	300	8.176	15	125
40	30	20	0.21	300	8.176	19	125
40	30	14	0.25	200	7.313	19	125
40	30	11	0.29	100	7.313	19	125
40	35	11	0.34	100	7.313	17	125
40	25	11	0.34	100	6.097	19	124
30	30	17	0.34	100	8.176	15	124
40	30	17	0.34	300	7.313	15	124
30	30	11	0.29	100	7.313	17	124
30	30	17	0.29	200	7.313	17	124
40	30	20	0.34	200	7.313	17	124
40	30	17	0.34	200	6.097	19	124
60	35	14	0.34	200	7.313	17	124
40	35	20	0.29	200	8.176	19	124
40	25	14	0.29	100	7.313	19	123
30	30	14	0.34	100	7.313	17	123
40	35	20	0.34	300	7.313	17	123
40	35	17	0.25	400	7.313	19	123
40	30	17	0.34	100	8.176	17	126
30	30	20	0.34	200	6.097	19	126
40	30	17	0.25	200	8.176	17	122
30	30	14	0.29	200	6.097	19	122
30	30	17	0.34	100	7.313	19	122
40	30	17	0.25	300	7.313	19	122
40	30	20	0.25	400	7.313	19	122
40	30	11	0.34	300	6.097	15	121
40	30	14	0.29	200	7.313	17	121
30	30	11	0.25	200	6.097	19	121
30	30	17	0.29	300	6.097	19	121
40	30	14	0.29	100	8.176	19	121
40	35	14	0.34	300	7.313	15	121
30	30	20	0.25	300	8.176	15	120
30	30	14	0.29	300	7.313	15	120
40	30	11	0.34	100	7.313	15	120
40	30	11	0.25	100	8.176	17	120
40	30	20	0.25	200	8.176	19	120
40	30	17	0.29	200	7.313	19	120
30	35	11	0.29	200	7.313	19	120
40	35	20	0.34	200	7.313	17	120
40	35	20	0.34	300	6.097	19	120

(continued)

Table 5.2 (continued)

B (m)	T (m)	H_c (m)	$\dfrac{c_u}{\sigma_v'}$	$\dfrac{E_{50}}{c_u}$	$\ln\left(\dfrac{EI}{\gamma_w h_{avg}^4}\right)$	γ (kN/m³)	δ_{hm} (mm)
30	25	11	0.25	100	7.313	17	122
30	30	20	0.34	300	7.313	15	122
30	30	17	0.34	300	6.097	17	118
40	30	14	0.21	200	8.176	19	118
40	25	20	0.34	100	7.313	19	117
30	30	20	0.29	200	8.176	15	117
40	30	17	0.29	300	7.313	17	117
50	30	20	0.34	200	7.313	19	117
50	25	20	0.25	300	7.313	17	116
50	25	11	0.34	100	7.313	17	116
40	30	20	0.29	300	8.176	15	116
30	30	20	0.29	300	7.313	17	116
30	30	20	0.29	200	7.313	19	116
40	30	17	0.21	300	8.176	19	116
40	30	17	0.34	100	8.176	19	116
40	25	20	0.34	200	6.097	19	115
30	30	14	0.25	200	8.176	15	115
30	30	20	0.34	100	8.176	17	115
40	30	11	0.29	100	8.176	17	115
40	30	14	0.34	100	8.176	17	115
60	30	20	0.29	200	8.176	17	115
30	30	17	0.29	100	8.176	19	115
30	30	11	0.29	200	6.097	17	118
30	30	14	0.34	200	6.097	17	118
30	30	14	0.34	100	8.176	15	114
40	30	14	0.29	100	8.176	17	114
40	25	20	0.29	200	8.176	17	114
30	35	14	0.34	100	8.176	19	114
40	25	20	0.25	200	7.313	19	113
40	25	17	0.29	300	6.097	19	113
50	30	20	0.34	200	8.176	15	113
50	30	14	0.21	300	8.176	17	113
40	30	17	0.34	200	7.313	17	113
30	30	20	0.34	300	6.097	19	113
30	30	20	0.25	300	8.176	19	113
40	30	20	0.29	300	8.176	19	113
40	30	20	0.29	200	7.313	17	112
40	25	17	0.29	200	7.313	15	112
30	30	11	0.29	200	7.313	15	112
30	30	14	0.34	200	8.176	15	112
40	30	14	0.25	300	8.176	19	112
40	35	14	0.29	200	8.176	17	112
60	30	14	0.25	300	7.313	19	111
30	30	17	0.25	200	7.313	19	111

(continued)

Table 5.2 (continued)

B (m)	T (m)	H_e (m)	$\dfrac{q_t}{\sigma_v}$	$\dfrac{E_{50}}{c_u}$	$\ln\left(\dfrac{EI}{\gamma_w h_{avg}^4}\right)$	γ (kN/m³)	δ_{hm} (mm)
40	30	20	0.21	200	8.846	19	115
50	35	14	0.34	200	7.313	17	115
40	35	14	0.25	300	7.313	19	111
30	25	20	0.25	200	7.313	17	110
30	25	20	0.34	100	7.313	17	110
40	25	17	0.25	200	7.313	19	110
30	30	17	0.29	200	8.176	15	110
40	30	11	0.21	200	8.176	17	110
50	35	11	0.29	200	7.313	17	110
40	35	14	0.25	200	8.176	19	110
40	35	17	0.25	300	8.176	19	110
40	35	14	0.29	300	6.097	19	110
40	25	17	0.34	200	6.097	19	109
40	30	14	0.25	200	8.176	17	109
30	30	17	0.25	300	7.313	19	109
40	30	17	0.25	400	7.313	19	109
40	35	11	0.25	300	6.097	19	109
40	30	17	0.29	300	8.176	15	108
30	30	17	0.34	100	8.176	17	108
40	30	14	0.29	300	6.097	19	108
40	35	11	0.34	100	7.313	19	108
30	25	17	0.25	200	7.313	17	107
30	30	11	0.34	100	6.097	19	111
40	30	17	0.25	200	8.176	19	111
30	30	11	0.29	100	7.313	19	107
40	30	17	0.21	200	8.846	19	107
40	35	11	0.29	200	8.176	15	107
40	35	14	0.29	200	7.313	19	107
30	35	17	0.29	300	7.313	19	107
40	25	11	0.21	200	7.313	17	106
50	25	11	0.25	200	7.313	17	106
40	30	17	0.34	200	8.176	15	106
30	30	14	0.29	200	7.313	17	106
40	30	17	0.25	300	8.176	17	106
40	30	17	0.29	200	8.176	17	106
40	30	14	0.34	300	6.097	17	106
30	30	20	0.25	200	8.176	19	106
40	30	11	0.34	100	7.313	19	106
40	30	20	0.34	200	7.313	19	106
40	30	17	0.29	300	7.313	17	105
30	30	17	0.29	200	8.176	17	105
40	30	17	0.29	200	7.313	19	105
40	30	11	0.29	100	8.176	19	105
30	30	14	0.34	100	8.176	19	105

(continued)

Table 5.2 (continued)

B (m)	T (m)	H_e (m)	$\dfrac{c_u}{\sigma'_v}$	$\dfrac{E_{50}}{c_u}$	$\ln\left(\dfrac{EI}{\gamma_w h^4_{avg}}\right)$	γ (kN/m³)	δ_{hm} (mm)
40	25	14	0.34	100	7.313	19	107
40	30	14	0.34	300	7.313	15	107
40	30	11	0.21	300	7.313	19	104
40	30	20	0.29	200	8.176	19	104
40	35	11	0.29	200	6.097	19	104
40	35	14	0.34	200	6.097	19	104
50	25	14	0.25	300	7.313	17	103
40	25	17	0.29	100	8.176	19	103
40	30	14	0.25	300	7.313	19	103
40	30	20	0.25	300	8.176	19	103
40	35	14	0.34	200	7.313	17	103
40	35	11	0.25	300	7.313	19	103
40	35	20	0.29	100	8.176	19	103
30	35	11	0.34	300	6.097	19	103
30	30	20	0.25	100	8.176	17	102
30	30	11	0.34	200	7.313	17	102
30	30	11	0.25	100	8.176	19	102
40	30	14	0.34	200	6.097	19	102
40	35	11	0.25	300	7.313	17	102
40	35	17	0.34	200	7.313	19	102
40	35	20	0.34	200	7.313	17	101
50	25	11	0.25	300	6.097	19	101
40	35	14	0.29	200	8.176	17	105
50	25	14	0.29	200	7.313	17	104
40	30	11	0.34	200	7.313	15	101
40	35	17	0.34	300	6.097	19	101
40	25	14	0.29	300	6.097	19	100
30	30	20	0.29	200	8.176	17	100
30	30	17	0.34	200	7.313	17	100
40	30	11	0.34	200	6.097	17	100
30	30	20	0.34	300	6.097	19	100
40	30	14	0.21	300	8.176	19	100
40	35	20	0.34	200	8.176	19	100
40	25	11	0.29	200	6.097	19	100
40	25	14	0.29	100	8.176	19	99
40	30	20	0.34	300	8.176	15	99
40	30	14	0.29	300	7.313	17	99
30	30	17	0.34	100	8.176	19	99
30	30	20	0.25	200	8.846	19	99
30	30	17	0.34	300	6.097	19	99
40	25	20	0.25	300	7.313	19	99
40	30	11	0.29	100	8.176	17	98
40	30	14	0.34	300	8.176	17	98
30	30	20	0.29	300	8.176	17	98

(continued)

Table 5.2 (continued)

B (m)	T (m)	H_e (m)	$\dfrac{q}{\sigma_v}$	$\dfrac{E_{50}}{c_u}$	$\ln\left(\dfrac{EI}{\gamma_w h_{avg}^4}\right)$	γ (kN/m³)	δ_{hm} (mm)
40	25	14	0.25	200	7.313	19	101
30	30	20	0.34	200	8.176	15	101
40	30	11	0.34	200	7.313	15	101
40	35	17	0.34	300	6.097	19	101
40	25	14	0.29	300	6.097	19	100
30	30	20	0.29	200	8.176	17	100
30	30	17	0.34	200	7.313	17	100
40	30	11	0.34	200	6.097	17	100
30	30	20	0.34	200	6.097	19	100
40	30	14	0.21	300	8.176	19	100
40	35	20	0.34	200	8.176	19	100
40	25	11	0.29	200	6.097	19	99
40	25	14	0.29	100	8.176	19	99
40	30	20	0.34	300	8.176	15	99
30	30	14	0.29	300	7.313	17	99
40	30	17	0.34	100	8.176	19	99
40	30	20	0.25	200	8.846	19	99
40	30	17	0.34	300	6.097	19	99
30	25	20	0.25	300	7.313	19	98
30	30	11	0.29	100	8.176	17	98
40	30	14	0.34	100	8.176	17	98
40	30	20	0.29	300	8.176	17	98
30	30	20	0.29	300	7.313	19	98
40	30	11	0.34	200	7.313	15	101
40	30	14	0.25	200	8.176	19	98
50	25	17	0.29	300	7.313	17	97
50	25	17	0.34	200	7.313	17	97
30	30	17	0.29	300	8.176	15	97
40	30	20	0.34	200	8.176	17	97
60	30	20	0.34	200	8.846	17	97
30	30	14	0.29	300	6.097	19	97
40	30	20	0.29	400	7.313	19	97
50	30	20	0.34	200	8.176	19	97
40	25	11	0.34	100	7.313	19	96
40	30	14	0.29	300	8.176	15	96
40	30	11	0.25	300	7.313	17	96
40	30	11	0.29	200	7.313	17	96
30	30	11	0.25	300	6.097	19	96
40	30	14	0.21	200	8.846	19	96
40	30	17	0.34	200	7.313	19	96
50	30	20	0.34	300	7.313	19	96
60	35	11	0.34	200	7.313	17	96
40	35	11	0.34	100	8.176	19	96
40	25	20	0.34	100	8.176	19	95

(continued)

Table 5.2 (continued)

B (m)	T (m)	H_e (m)	$\frac{c_u}{\sigma'_v}$	$\frac{E_{50}}{c_u}$	$\ln\left(\frac{EI}{\gamma_w h^4_{avg}}\right)$	γ (kN/m³)	δ_{hm} (mm)
30	30	20	0.29	300	7.313	19	98
40	30	11	0.21	200	8.176	19	98
30	30	14	0.34	300	6.097	17	95
60	30	20	0.29	200	8.846	17	95
40	30	17	0.29	200	8.176	19	95
30	25	17	0.25	300	7.313	17	94
40	25	17	0.25	300	7.313	19	94
40	25	17	0.29	200	7.313	19	94
30	30	17	0.34	200	8.176	15	94
40	30	14	0.34	200	8.176	15	94
40	30	17	0.34	300	7.313	17	94
30	30	17	0.29	200	8.176	17	93
30	30	20	0.34	300	7.313	17	93
40	30	11	0.21	300	8.176	17	93
60	30	11	0.34	200	7.313	17	93
30	30	20	0.34	200	7.313	19	93
40	35	20	0.34	300	7.313	19	93
40	25	17	0.34	100	8.176	19	92
40	30	17	0.34	300	8.176	19	92
40	30	14	0.25	300	8.176	15	92
40	35	14	0.25	200	8.846	19	92
30	25	11	0.34	100	7.313	17	91
30	30	14	0.34	300	7.313	15	95
30	30	17	0.25	300	8.176	17	95
30	30	14	0.25	300	7.313	19	91
30	30	11	0.29	200	6.097	19	91
40	30	11	0.34	100	8.176	19	91
50	35	11	0.29	200	8.176	17	91
40	35	17	0.29	200	8.846	19	91
40	30	11	0.25	200	8.176	17	91
60	30	11	0.29	300	7.313	17	90
60	30	17	0.34	200	8.846	17	90
30	30	14	0.34	200	6.097	19	90
40	30	14	0.25	400	7.313	19	90
50	35	14	0.34	300	7.313	17	90
30	30	20	0.34	300	8.176	15	89
40	30	17	0.29	300	8.176	17	89
40	30	17	0.34	200	8.176	17	89
30	30	11	0.34	100	7.313	19	89
30	30	14	0.34	100	8.176	19	89
30	25	11	0.25	200	7.313	19	88
50	25	14	0.29	300	7.313	17	88
40	25	14	0.34	100	8.176	19	88
40	25	17	0.34	300	6.097	19	88

(continued)

Table 5.2 (continued)

B (m)	T (m)	H_e (m)	$\frac{\sigma_h}{\sigma_v}$	$\frac{E_{50}}{c_u}$	$\ln\left(\frac{EI}{\gamma_w h^4_{avg}}\right)$	γ (kN/m³)	δ_{hm} (mm)
50	25	11	0.29	200	7.313	17	91
60	30	14	0.29	200	8.176	17	91
50	35	11	0.34	200	7.313	17	88
40	35	14	0.29	200	8.176	19	88
40	35	17	0.29	300	8.176	19	88
50	25	17	0.29	200	8.176	17	87
40	25	11	0.25	200	7.313	19	87
30	30	20	0.29	300	8.176	17	87
60	30	17	0.29	200	8.846	17	87
30	30	17	0.34	300	6.097	19	87
40	30	20	0.29	200	8.846	19	87
40	30	20	0.29	300	8.176	19	87
40	30	20	0.34	300	7.313	19	87
40	35	11	0.34	300	7.313	15	87
30	25	14	0.25	300	7.313	17	86
30	30	14	0.29	300	8.176	15	86
30	30	11	0.25	200	7.313	19	86
30	30	20	0.25	200	8.846	19	86
40	30	17	0.29	400	7.313	19	86
40	35	17	0.34	200	8.176	19	86
30	25	14	0.29	200	7.313	17	85
30	30	11	0.34	100	8.176	17	85
30	30	11	0.34	200	6.097	17	88
30	30	14	0.29	200	7.313	19	88
40	25	20	0.34	200	7.313	19	84
30	30	14	0.34	200	7.313	17	84
40	35	11	0.25	200	8.176	19	84
30	30	11	0.29	200	8.176	15	83
30	30	14	0.34	200	8.176	15	83
30	30	11	0.29	200	7.313	17	83
30	30	17	0.25	300	8.176	19	83
30	30	17	0.34	200	7.313	19	83
40	30	14	0.29	200	8.176	19	83
30	25	20	0.34	200	7.313	17	82
50	25	17	0.34	300	7.313	17	82
30	30	17	0.34	300	8.176	15	82
40	30	20	0.34	300	8.176	17	82
30	30	17	0.29	200	8.176	19	82
50	25	14	0.29	200	8.176	17	81
40	30	11	0.21	200	8.846	19	81
40	30	14	0.25	200	8.846	19	81
40	30	11	0.29	200	7.313	19	81
30	30	14	0.29	300	7.313	19	81
30	30	14	0.34	200	7.313	19	81

(continued)

Table 5.2 (continued)

B (m)	T (m)	H_e (m)	$\frac{\sigma_h}{\sigma_v}$	$\frac{E_{50}}{c_u}$	$\ln\left(\frac{EI}{\gamma_w h^4_{avg}}\right)$	γ (kN/m³)	δ_{hm} (mm)
30	30	20	0.34	200	8.176	17	85
30	30	14	0.25	200	8.176	19	85
40	35	11	0.29	300	6.097	19	81
40	35	14	0.34	300	6.097	19	81
40	35	20	0.34	300	8.176	19	81
40	25	17	0.29	300	7.313	19	80
40	25	11	0.34	100	8.176	15	80
40	30	14	0.34	300	8.176	17	80
60	30	14	0.34	200	8.846	19	80
30	30	17	0.25	200	8.846	19	80
40	30	14	0.25	300	8.176	19	80
40	30	17	0.29	200	8.846	19	80
40	30	14	0.34	300	6.097	19	79
40	30	11	0.34	200	8.176	15	79
30	30	17	0.29	300	8.176	17	79
40	30	11	0.34	300	6.097	17	79
40	30	11	0.34	200	6.097	19	79
40	30	17	0.34	200	8.176	19	79
40	35	11	0.34	200	6.097	19	78
40	25	20	0.29	200	8.176	19	78
40	25	11	0.34	200	6.097	19	78
30	30	17	0.34	200	8.176	17	78
40	35	11	0.29	200	8.176	17	81
40	35	11	0.29	200	7.313	19	81
40	35	11	0.34	200	7.313	17	78
40	25	14	0.34	300	6.097	19	77
40	30	11	0.29	200	8.176	17	77
40	30	14	0.29	300	8.176	17	77
30	30	20	0.29	300	8.176	19	77
30	30	11	0.34	100	8.176	19	77
30	30	20	0.34	300	7.313	19	77
40	25	20	0.25	300	8.176	19	76
40	30	11	0.29	300	7.313	17	76
40	30	11	0.34	200	7.313	17	76
40	30	17	0.29	400	7.313	19	76
30	35	14	0.29	200	8.846	19	76
40	25	17	0.29	200	8.176	19	75
30	30	20	0.34	200	8.176	19	75
40	30	20	0.34	400	7.313	19	75
30	25	11	0.25	300	7.313	17	74
50	25	14	0.34	300	7.313	17	74
40	30	11	0.25	300	8.176	17	74
40	30	17	0.34	300	8.176	17	74
40	25	14	0.34	200	7.313	19	73

(continued)

Table 5.2 (continued)

B (m)	T (m)	H_e (m)	$\frac{\sigma_h}{\sigma_v}$	$\frac{E_{50}}{c_u}$	$\ln\left(\frac{EI}{\gamma_w h_{avg}^4}\right)$	γ (kN/m³)	δ_{hm} (mm)	B (m)	T (m)	H_e (m)	$\frac{\sigma_h}{\sigma_v}$	$\frac{E_{50}}{c_u}$	$\ln\left(\frac{EI}{\gamma_w h_{avg}^4}\right)$	γ (kN/m³)	δ_{hm} (mm)
40	30	11	0.25	300	7.313	19	78	60	30	11	0.29	200	8.176	17	73
40	30	17	0.29	300	8.176	19	78	30	25	20	0.29	200	8.176	17	72
50	25	11	0.29	200	8.176	17	72	30	30	11	0.34	200	6.097	19	69
40	25	17	0.25	300	8.176	19	72	30	30	14	0.34	200	7.313	19	69
30	30	20	0.34	300	8.176	17	72	40	25	20	0.34	200	8.176	19	68
40	30	20	0.34	300	8.176	19	72	30	30	14	0.34	300	7.313	17	68
40	35	11	0.25	200	8.846	19	72	60	30	11	0.34	200	8.846	17	68
40	25	11	0.25	300	7.313	19	71	30	30	17	0.34	200	8.176	19	68
40	25	20	0.34	300	7.313	19	71	40	30	14	0.34	200	7.313	19	68
30	30	14	0.34	300	8.176	15	71	40	30	17	0.34	400	8.176	19	68
30	30	14	0.25	300	8.176	19	71	30	35	17	0.34	300	8.176	19	68
30	30	14	0.29	300	7.313	19	71	30	30	11	0.29	300	7.313	17	67
30	30	14	0.34	300	6.097	19	71	30	30	14	0.29	300	8.176	17	67
40	30	14	0.29	200	8.846	19	71	30	30	17	0.34	300	7.313	19	67
30	25	17	0.29	200	8.176	17	70	40	30	20	0.34	400	7.313	19	67
30	25	20	0.34	300	7.313	17	70	40	30	11	0.25	200	8.846	19	67
30	30	14	0.25	200	8.846	19	70	40	30	11	0.29	200	8.176	19	67
40	35	14	0.29	300	8.176	19	70	50	35	11	0.34	300	7.313	17	67
40	25	14	0.29	200	8.176	19	69	30	25	17	0.34	300	7.313	17	66
30	30	11	0.34	300	6.097	17	69	40	25	20	0.29	300	8.176	19	66
30	30	11	0.25	300	7.313	19	69	40	25	17	0.34	300	7.313	19	66
30	30	11	0.29	200	7.313	19	69	30	30	11	0.29	200	8.176	17	66

(continued)

Table 5.2 (continued)

B (m)	T (m)	H_e (m)	$\dfrac{c_u}{\sigma'_v}$	$\dfrac{E_{50}}{c_u}$	$\ln\left(\dfrac{EI}{\gamma_w h^4_{avg}}\right)$	γ (kN/m³)	δ_{hm} (mm)
30	30	14	0.29	200	8.846	19	69
30	30	17	0.29	300	8.176	19	69
30	25	14	0.29	200	8.176	17	65
30	30	11	0.25	300	8.176	17	65
30	30	17	0.34	300	8.176	17	65
30	30	20	0.34	200	8.846	19	64
40	30	17	0.34	300	8.176	19	64
40	35	11	0.34	200	7.313	19	64
40	35	14	0.34	300	7.313	19	64
40	30	11	0.34	200	8.176	17	63
40	30	14	0.34	300	8.176	17	63
30	30	20	0.34	300	8.176	19	63
40	30	14	0.34	300	7.313	19	63
60	30	11	0.29	200	8.846	17	62
40	40	11	0.29	300	7.313	19	62
30	30	11	0.34	200	7.313	17	61
50	25	11	0.34	300	7.313	17	61
40	25	11	0.34	200	7.313	19	61
30	25	11	0.29	400	7.313	19	61
30	30	14	0.29	400	8.176	19	61
40	30	17	0.34	300	6.097	19	61
40	35	11	0.34	300	6.097	19	61
40	30	14	0.29	300	8.176	19	66
40	35	11	0.25	300	8.176	19	66
40	30	11	0.34	300	7.313	17	60
40	30	11	0.29	200	8.846	19	60
40	30	14	0.34	400	7.313	19	60
40	25	11	0.29	300	7.313	19	58
40	25	14	0.29	300	8.176	19	58
40	25	14	0.34	300	7.313	19	58
30	30	14	0.34	300	8.176	19	58
30	30	17	0.34	200	8.846	19	58
30	30	17	0.34	400	7.313	19	58
40	25	20	0.34	300	8.176	19	57
30	30	14	0.29	300	8.176	19	57
40	35	11	0.34	200	8.176	19	57
30	30	11	0.29	200	8.176	19	56
40	30	11	0.34	200	8.176	19	56
30	30	11	0.25	300	8.176	19	55
40	30	11	0.34	300	8.176	19	55
30	30	17	0.34	300	8.176	19	54
30	25	17	0.34	200	8.176	17	54
40	30	11	0.29	300	7.313	19	54
40	30	11	0.34	200	7.313	19	54

(continued)

Table 5.2 (continued)

B (m)	T (m)	H_e (m)	$\dfrac{\sigma_h}{\sigma_v}$	$\dfrac{E_{50}}{c_u}$	$\ln\!\left(\dfrac{EI}{\gamma_w h_{avg}^4}\right)$	γ (kN/m³)	δ_{hm} (mm)
30	25	14	0.34	300	7.313	17	60
40	25	11	0.34	300	6.097	19	60
40	30	14	0.34	300	8.176	19	53
40	25	11	0.34	200	8.176	19	52
40	35	11	0.29	300	8.176	19	52
30	30	11	0.34	300	7.313	17	51
40	30	11	0.29	300	8.176	19	51
30	30	14	0.34	200	8.846	19	50
40	30	11	0.34	400	7.313	19	50
30	25	11	0.34	300	7.313	17	49
30	30	11	0.29	200	8.846	19	49
40	30	11	0.34	300	7.313	19	49
40	35	11	0.34	300	7.313	19	49
40	25	14	0.34	300	8.176	19	48
40	25	11	0.34	300	7.313	19	47

B (m)	T (m)	H_e (m)	$\dfrac{\sigma_h}{\sigma_v}$	$\dfrac{E_{50}}{c_u}$	$\ln\!\left(\dfrac{EI}{\gamma_w h_{avg}^4}\right)$	γ (kN/m³)	δ_{hm} (mm)
40	35	14	0.34	300	8.176	19	54
30	30	11	0.34	300	6.097	19	53
30	30	11	0.34	200	8.176	19	47
30	30	14	0.34	400	7.313	19	46
30	30	11	0.29	400	7.313	19	45
30	30	14	0.34	300	8.176	19	45
30	30	11	0.34	300	8.176	17	42
30	30	11	0.34	200	8.846	19	42
30	35	11	0.34	300	7.313	19	42
30	30	11	0.34	300	8.176	19	42
30	30	17	0.34	400	8.846	19	41
40	25	11	0.34	300	8.176	19	40
30	30	11	0.34	400	7.313	19	35
30	30	14	0.34	400	8.846	19	34
30	30	11	0.34	400	8.846	19	26

Table 5.3 A summary of diaphragm wall deflection testing data set

B (m)	T (m)	H_e (m)	$\frac{c_u}{\sigma_v'}$	$\frac{E_{50}}{c_u}$	$\ln\left(\frac{EI}{\gamma_w h_{avg}^4}\right)$	γ (kN/m³)	δ_{hm} (mm)
30	30	17	0.25	100	6.097	15	412
40	30	20	0.25	200	6.097	15	383
30	30	14	0.25	100	6.097	15	373
30	30	20	0.25	200	6.097	15	358
50	30	20	0.25	100	6.097	19	345
50	35	20	0.25	100	7.313	17	334
40	30	20	0.34	100	6.097	15	324
40	30	14	0.25	100	6.097	17	313
50	35	20	0.29	200	6.097	17	307
40	35	20	0.25	200	7.313	15	304
40	35	14	0.25	100	7.313	15	300
30	30	20	0.25	100	6.097	19	294
40	30	17	0.29	100	6.097	17	288
30	30	20	0.29	100	6.097	17	282
30	30	17	0.29	200	6.097	15	273
30	30	20	0.25	200	6.097	17	271
40	35	17	0.29	100	6.097	19	268
30	30	17	0.29	100	6.097	17	262
40	30	17	0.29	300	6.097	15	260
40	30	20	0.25	300	6.097	17	257
40	35	11	0.25	100	7.313	15	243
40	30	17	0.29	100	6.097	19	240
30	30	20	0.25	100	7.313	15	238
30	30	14	0.29	200	6.097	15	234
30	30	20	0.34	200	6.097	15	233
50	35	17	0.25	200	7.313	17	225
40	35	20	0.29	100	7.313	19	220
50	30	20	0.25	300	6.097	19	218
40	30	17	0.34	200	6.097	15	232
50	35	20	0.34	100	7.313	17	230
50	35	20	0.25	100	7.313	19	228
30	30	17	0.29	100	6.097	19	215
40	30	20	0.25	200	7.313	15	211
30	30	14	0.25	100	7.313	15	210
40	35	20	0.34	100	7.313	17	210
30	30	20	0.29	100	7.313	15	206
50	35	17	0.34	100	7.313	17	206
30	30	14	0.29	300	6.097	15	204
40	35	20	0.25	100	8.176	19	204
40	30	14	0.29	100	7.313	15	201

(continued)

Table 5.3 (continued)

B (m)	T (m)	H_e (m)	$\frac{c_u}{\sigma_v'}$	$\frac{E_{50}}{c_u}$	$\ln\left(\frac{EI}{\gamma_w h_{avg}^4}\right)$	γ (kN/m³)	δ_{hm} (mm)
40	30	17	0.25	100	7.313	15	251
40	30	14	0.29	100	6.097	17	250
30	30	17	0.29	100	7.313	15	197
40	30	17	0.25	100	7.313	19	197
40	30	14	0.25	300	6.097	17	195
30	30	20	0.25	200	7.313	15	193
40	35	14	0.25	200	6.097	19	191
40	25	17	0.29	100	6.097	19	188
40	30	20	0.25	100	8.176	15	187
30	30	20	0.25	100	7.313	19	186
30	30	20	0.25	300	6.097	19	184
40	30	20	0.29	200	7.313	15	183
30	30	11	0.25	100	6.097	19	181
40	30	17	0.25	100	8.176	15	180
30	30	11	0.29	200	6.097	15	179
40	35	17	0.34	200	7.313	15	179
40	35	11	0.25	200	7.313	15	178
30	30	20	0.34	100	7.313	17	177
40	35	17	0.25	300	7.313	15	177
40	35	20	0.34	300	7.313	17	176
60	35	20	0.34	200	7.313	17	175
30	30	17	0.25	100	7.313	19	174
40	30	17	0.25	200	7.313	15	200
40	30	20	0.34	100	7.313	15	198
40	30	20	0.21	300	7.313	19	170
40	35	20	0.29	300	6.097	19	169
60	30	14	0.25	200	7.313	17	168
30	30	17	0.34	100	6.097	19	167
40	30	20	0.29	100	8.176	15	166
40	35	17	0.29	200	7.313	17	166
50	30	20	0.29	300	6.097	19	165
40	35	11	0.29	100	6.097	19	164
50	25	14	0.25	100	7.313	17	162
40	30	11	0.29	100	6.097	19	162
60	30	17	0.29	200	7.313	17	161
40	30	17	0.29	100	8.176	15	160
40	35	14	0.25	100	8.176	19	160
30	30	20	0.29	100	7.313	19	158
40	30	11	0.34	100	6.097	17	157
50	35	20	0.29	200	8.176	17	157
40	30	17	0.21	300	7.313	19	156
40	35	20	0.29	200	7.313	19	155
60	30	20	0.21	200	8.176	17	153
30	30	14	0.34	100	7.313	15	152

(continued)

Table 5.3 (continued)

B (m)	T (m)	H_e (m)	$\frac{c_u}{\sigma'_v}$	$\frac{E_{50}}{c_u}$	$\ln\left(\frac{EI}{\gamma_w h^4_{avg}}\right)$	γ (kN/m³)	δ_{hm} (mm)
40	30	17	0.29	200	7.313	15	172
50	30	20	0.25	200	7.313	19	171
30	25	20	0.29	200	6.097	17	150
30	30	17	0.25	200	7.313	17	150
60	30	20	0.29	300	7.313	17	150
30	30	20	0.34	100	7.313	19	149
40	30	14	0.29	100	7.313	15	149
40	30	20	0.25	200	8.176	15	148
30	30	17	0.29	100	7.313	19	147
40	35	20	0.25	400	7.313	19	147
50	35	14	0.29	200	7.313	17	146
40	30	14	0.29	300	6.097	17	145
40	30	17	0.21	200	8.176	17	144
40	30	17	0.29	100	8.176	17	143
30	30	17	0.29	100	8.176	15	142
40	30	11	0.29	100	7.313	17	142
40	35	14	0.25	300	7.313	17	142
30	30	11	0.29	100	6.097	19	141
40	35	17	0.25	300	7.313	19	141
40	30	17	0.29	200	7.313	17	140
40	30	20	0.34	200	6.097	19	140
30	30	17	0.34	100	7.313	17	139
40	35	14	0.21	200	8.846	15	152
50	25	17	0.29	100	7.313	17	151
50	25	20	0.34	100	7.313	17	136
60	30	17	0.29	300	7.313	17	136
40	30	11	0.25	200	6.097	19	135
40	35	14	0.29	100	8.176	19	135
40	30	14	0.21	300	7.313	19	134
30	25	14	0.29	200	6.097	17	133
60	30	17	0.34	200	7.313	17	133
40	35	17	0.29	200	7.313	19	133
40	30	20	0.25	300	8.176	15	132
40	30	14	0.34	200	6.097	17	132
40	35	17	0.25	200	8.176	19	132
40	30	14	0.34	100	8.176	15	131
30	30	11	0.25	100	8.176	17	131
40	30	17	0.25	200	8.176	15	127
40	25	17	0.21	200	7.313	17	126
30	30	11	0.21	200	7.313	19	126
40	30	17	0.29	100	8.176	17	125
40	30	14	0.34	100	7.313	19	125
40	30	17	0.29	200	8.176	15	124
30	30	14	0.25	300	7.313	17	124

(continued)

Table 5.3 (continued)

B (m)	T (m)	H_e (m)	$\dfrac{c_u}{\sigma_v'}$	$\dfrac{E_{50}}{c_u}$	$\ln\left(\dfrac{EI}{\gamma_w h_{avg}^4}\right)$	γ (kN/m³)	δ_{hm} (mm)	B (m)	T (m)	H_e (m)	$\dfrac{c_u}{\sigma_v'}$	$\dfrac{E_{50}}{c_u}$	$\ln\left(\dfrac{EI}{\gamma_w h_{avg}^4}\right)$	γ (kN/m³)	δ_{hm} (mm)
40	30	14	0.29	200	6.097	19	138	40	35	20	0.25	200	8.846	19	124
30	30	17	0.29	300	7.313	15	137	40	30	20	0.34	100	8.176	19	123
40	25	14	0.29	200	6.097	19	122	50	25	17	0.25	300	7.313	17	112
30	30	20	0.29	100	8.176	19	122	30	30	17	0.34	300	7.313	15	112
50	25	14	0.25	200	7.313	17	121	40	35	11	0.25	100	8.176	19	131
40	25	11	0.29	100	7.313	19	111	40	25	17	0.29	100	7.313	19	130
30	30	17	0.34	200	6.097	19	111	40	30	20	0.25	200	8.176	17	130
30	25	11	0.29	200	6.097	17	110	40	30	20	0.29	300	7.313	17	129
30	30	20	0.34	200	7.313	17	110	30	25	20	0.21	300	7.313	17	128
40	35	17	0.25	200	8.846	19	110	40	35	17	0.34	200	7.313	17	128
30	30	17	0.25	200	8.176	17	109	30	30	20	0.25	300	7.313	19	121
40	30	20	0.29	300	7.313	19	109	30	30	14	0.21	200	7.313	17	120
40	30	11	0.25	300	6.097	19	108	30	25	14	0.25	100	8.176	19	120
30	25	17	0.34	100	7.313	17	107	50	30	20	0.29	300	7.313	19	120
30	30	14	0.34	100	7.313	19	107	40	25	20	0.29	300	6.097	19	119
40	35	17	0.29	200	8.176	19	107	40	30	11	0.25	200	7.313	17	118
40	25	20	0.29	100	8.176	19	106	40	30	20	0.25	200	8.176	17	117
50	35	14	0.29	200	8.176	17	116	50	25	20	0.29	200	7.313	17	116
40	30	20	0.25	300	8.176	17	115	40	30	14	0.29	300	7.313	17	116
60	30	14	0.34	200	7.313	17	115	60	30	11	0.29	300	6.097	17	106
40	25	17	0.34	100	7.313	19	114	40	25	20	0.34	100	8.176	19	106
50	35	17	0.34	300	7.313	17	114	50	35	20	0.34	300	7.313	17	105

(continued)

Table 5.3 (continued)

B (m)	T (m)	H_e (m)	$\frac{c_u}{\sigma_v}$	$\frac{E_{50}}{c_u}$	$\ln\left(\frac{EI}{\gamma_w h^4_{avg}}\right)$	γ (kN/m³)	δ_{hm} (mm)
30	30	17	0.25	300	8.176	15	113
60	30	11	0.29	200	7.313	17	113
40	35	20	0.29	200	8.846	19	104
30	30	14	0.29	100	8.176	19	103
40	35	17	0.34	300	7.313	17	103
30	25	14	0.34	100	7.313	17	102
40	30	14	0.29	200	7.313	19	102
50	25	20	0.29	300	7.313	17	101
30	30	14	0.25	300	8.176	15	101
30	25	14	0.25	200	7.313	17	100
40	30	11	0.34	100	8.176	17	100
40	35	11	0.29	200	7.313	17	100
30	30	14	0.29	200	8.176	15	99
40	30	11	0.25	200	7.313	19	99
40	25	20	0.29	200	7.313	19	98
30	30	17	0.25	200	8.176	19	98
30	25	20	0.25	300	7.313	17	97
30	30	14	0.25	200	8.176	17	97
40	30	17	0.29	300	7.313	19	97
40	30	14	0.34	200	8.176	19	96
40	25	14	0.34	200	6.097	17	96
50	30	20	0.29	300	7.313	19	96
40	30	11	0.29	200	6.097	19	105
30	30	20	0.29	300	8.176	15	104
40	35	14	0.25	400	7.313	19	95
40	25	20	0.34	300	6.097	19	94
40	30	17	0.25	300	8.176	19	94
40	30	14	0.29	200	8.176	17	93
30	25	17	0.29	200	7.313	17	92
30	30	20	0.25	300	8.176	19	92
40	25	11	0.29	100	8.176	19	91
40	30	17	0.25	200	8.846	19	91
50	25	20	0.29	200	8.176	17	90
30	30	20	0.29	200	8.176	19	90
50	25	14	0.34	200	7.313	17	89
30	30	11	0.29	100	8.176	19	89
50	25	11	0.25	300	7.313	17	88
30	30	14	0.29	300	7.313	17	88
40	35	14	0.25	300	8.176	19	88
50	25	20	0.34	300	7.313	17	87
30	30	17	0.29	300	7.313	19	87
40	30	20	0.34	200	8.176	19	87
40	25	14	0.29	200	7.313	19	86
30	30	20	0.29	400	7.313	19	86

(continued)

Table 5.3 (continued)

B (m)	T (m)	H_e (m)	$\dfrac{c_u}{\sigma_v'}$	$\dfrac{E_{50}}{c_u}$	$\ln\left(\dfrac{EI}{\gamma_w h_{avg}^4}\right)$	γ (kN/m³)	δ_{hm} (mm)
30	25	20	0.29	200	7.313	17	95
30	30	11	0.29	300	6.097	17	95
40	35	14	0.29	300	7.313	19	84
30	30	17	0.34	300	7.313	17	83
40	35	14	0.34	200	7.313	19	83
30	30	14	0.25	300	8.176	17	82
30	30	14	0.29	200	8.176	17	81
40	30	11	0.29	300	6.097	19	81
40	35	14	0.34	300	7.313	17	81
40	35	17	0.34	300	7.313	19	81
40	25	17	0.34	200	7.313	19	80
40	30	11	0.21	300	8.176	19	80
40	35	11	0.25	300	7.313	19	80
40	30	11	0.25	200	8.176	19	79
30	25	17	0.34	200	7.313	17	78
40	30	14	0.34	300	7.313	17	78
50	25	11	0.34	200	7.313	17	77
40	30	14	0.34	200	8.176	17	77
40	30	17	0.34	300	7.313	19	77
60	30	14	0.29	200	8.846	17	76
30	30	20	0.29	200	8.846	19	75
50	25	11	0.29	300	7.313	17	74
40	25	14	0.25	300	7.313	19	85
40	25	20	0.29	300	7.313	19	84
30	30	11	0.29	300	6.097	19	72
40	25	14	0.29	300	7.313	19	71
30	30	14	0.29	200	8.176	19	71
40	35	14	0.34	200	8.176	19	71
40	30	14	0.29	400	7.313	19	70
30	30	11	0.25	200	8.176	19	69
30	30	17	0.29	200	8.846	19	69
40	30	11	0.29	200	8.176	19	69
40	30	20	0.29	400	8.176	19	68
40	35	11	0.25	400	7.313	19	68
40	30	14	0.34	200	8.176	17	67
40	30	11	0.25	400	7.313	19	67
30	25	14	0.25	300	8.176	19	66
40	30	11	0.34	200	7.313	17	66
50	25	17	0.34	200	8.176	19	65
40	30	11	0.34	200	7.313	19	64
40	25	17	0.29	300	8.176	19	63
60	30	11	0.25	300	8.176	19	63
30	35	11	0.29	300	7.313	19	62
40	30	11	0.29	300	8.176	17	61

(continued)

Table 5.3 (continued)

B (m)	T (m)	H_e (m)	$\frac{c_u}{\sigma_v'}$	$\frac{E_{50}}{c_u}$	$\ln\left(\frac{EI}{\gamma_w h_{avg}^4}\right)$	γ (kN/m³)	δ_{hm} (mm)
30	25	11	0.29	200	7.313	17	73
30	25	14	0.34	200	7.313	17	72
40	25	14	0.34	200	8.176	19	60
40	35	11	0.34	300	7.313	17	60
30	30	11	0.25	200	8.846	19	58
30	25	11	0.29	200	8.176	17	57
40	25	11	0.25	300	8.176	19	56
30	30	14	0.34	300	7.313	19	55
30	30	14	0.34	300	8.176	17	54
30	30	11	0.29	300	8.176	17	53

B (m)	T (m)	H_e (m)	$\frac{c_u}{\sigma_v'}$	$\frac{E_{50}}{c_u}$	$\ln\left(\frac{EI}{\gamma_w h_{avg}^4}\right)$	γ (kN/m³)	δ_{hm} (mm)
40	35	11	0.29	200	8.846	19	61
40	25	14	0.34	200	8.176	19	60
40	30	11	0.29	400	7.313	19	52
30	30	14	0.29	400	8.176	19	50
40	30	11	0.34	300	8.176	17	49
40	25	11	0.29	300	8.176	19	48
30	30	20	0.34	400	8.846	19	47
30	30	11	0.29	300	8.176	19	44
40	30	11	0.34	300	8.176	19	42
30	30	11	0.29	400	8.176	19	37

Fig. 5.2 Performance of MARS model for diaphragm wall deflection: **a** training and **b** testing

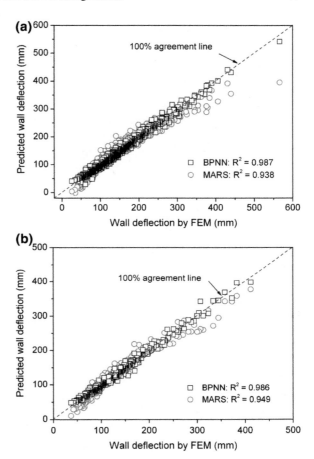

Table 5.4 Comparison of performance measures for BPNN and MARS

Methods	Processing time (s)	Accuracy					
		Training sets			Testing sets		
		R^2	MSE	MAE	R^2	MSE	MAE
MARS	1.11	0.938	303.6	12.21	0.949	233.2	11.43
BPNN	6.31	0.987	65.69	6.059	0.986	61.84	5.715

5.4 Model Interpretability

Table 5.5 lists the BFs and their corresponding equations for the optimal MARS model. The interpretable MARS model is given by

$$
\begin{aligned}
\delta_{hm\_\mathrm{MARS}} = {}& 165 - 50.889 \times \mathrm{BF1} + 66.598 \times \mathrm{BF2} - 0.1914 \times \mathrm{BF3} + 0.4956 \times \mathrm{BF4} \\
& - 10.324 \times \mathrm{BF5} + 19.135 \times \mathrm{BF6} - 326.34 \times \mathrm{BF7} + 815.69 \times \mathrm{BF8} + 4.9881 \times \mathrm{BF9} \\
& - 6.1891 \times \mathrm{BF10} + 7.4897 \times \mathrm{BF11} - 7.0073 \times \mathrm{BF12} - 13.712 \times \mathrm{BF13} + 24.131 \times \mathrm{BF14} \\
& + 540.93 \times \mathrm{BF15} - 331.28 \times \mathrm{BF16} + 2.7716 \times \mathrm{BF17} - 4.5821 \times \mathrm{BF18} - 1.1808 \times \mathrm{BF19} \\
& + 0.8612 \times \mathrm{BF20} + 0.5114 \times \mathrm{BF21} - 1.5474 \times \mathrm{BF22}
\end{aligned}
\tag{5.1}
$$

Table 5.5 Basis functions and corresponding equations of MARS model for diaphragm wall deflection prediction

Basis function	Equation
BF1	$\max(0,\ \ln(\mathrm{EI}/\gamma_w h_{\mathrm{avg}}^4) - 7.313)$
BF2	$\max(0,\ 7.313 - \ln(\mathrm{EI}/\gamma_w h_{\mathrm{avg}}^4))$
BF3	$\max(0,\ E_{50}/c_u - 200)$
BF4	$\max(0,\ 200 - E_{50}/c_u)$
BF5	$\max(0,\ \gamma - 17)$
BF6	$\max(0,\ 17 - \gamma)$
BF7	$\max(0,\ c_u/\gamma'_v - 0.25)$
BF8	$\max(0,\ 0.25 - c_u/\gamma'_v)$
BF9	$\max(0,\ h - 17)$
BF10	$\max(0,\ 17 - h)$
BF11	$\max(0,\ T - 30)$
BF12	$\max(0,\ 30 - T)$
BF13	$\mathrm{BF6} \times \max(0,\ \ln(\mathrm{EI}/\gamma_w h_{\mathrm{avg}}^4) - 7.313)$
BF14	$\mathrm{BF6} \times \max(0,\ 7.313 - \ln(\mathrm{EI}/\gamma_w h_{\mathrm{avg}}^4))$
BF15	$\mathrm{BF7} \times \max(0,\ \ln(\mathrm{EI}/\gamma_w h_{\mathrm{avg}}^4) - 8.176)$
BF16	$\mathrm{BF7} \times \max(0,\ 8.176 - \ln(\mathrm{EI}/\gamma_w h_{\mathrm{avg}}^4))$
BF17	$\mathrm{BF10} \times \max(0,\ \ln(\mathrm{EI}/\gamma_w h_{\mathrm{avg}}^4) - 7.313)$
BF18	$\mathrm{BF10} \times \max(0,\ 7.313 - \ln(\mathrm{EI}/\gamma_w h_{\mathrm{avg}}^4))$
BF19	$\mathrm{BF10} \times \max(0,\ T - 30)$
BF20	$\mathrm{BF10} \times \max(0,\ 30 - T)$
BF21	$\max(0,\ B - 40)$
BF22	$\max(0,\ 40 - B)$

5.5 Parameter Relative Importance

The ANOVA parameter relative importance assessment indicates that the two variables which contribute most to the diaphragm wall deflection are h (excavation depth) and B (excavation width). For brevity, the ANOVA decomposition has been omitted.

5.6 Assessment of the Proposed MARS Model

It should be admitted that previously all the results are based on numerical simulation. Thus, the applicability of it should be validated through the documented case histories.

In all the previous analyses, the groundwater table was assumed at the ground surface, which is the most unfavorable condition. In many situations with soft clay, the water could be 1–2 m below the ground surface. Additional analyses carried out to investigate the influence of the groundwater table indicate that the maximum wall deflection decreases almost linearly with decreasing groundwater level. For brevity, these plots have been omitted. The water table correction factor μ_w can be approximated as $\mu_w = 1 - 0.1l$, where l is the depth of the groundwater table below the ground surface (in meters) and $l \leq 2$. Thus, the predicted maximum wall deflection $\delta^*_{hm\_MARS}$ can be estimated using

$$\delta^*_{hm\_MARS} = \mu_w \delta_{hm\_MARS} \tag{5.2}$$

To validate the proposed MARS model, a total of 21 well-documented excavation case histories from various countries as listed in Table 5.6 were analyzed. Also listed in Table 5.6 are the predicted $\delta^*_{hm\_MARS}$ values. Figure 5.3 shows the predicted wall deflections versus the measured for the 21 cases. The plot indicates that MARS model is able to predict reasonably well (within a factor of 2) the excavation-induced wall deflections for the case histories considered.

5.7 Summary

Based on an extensive number of finite element numerical simulations, both MARS and BPNN models were developed for relating the maximum wall deflection induced by braced excavation to the influential factors including the excavation geometries, the unit weight, the system stiffness, the soil strength, and the stiffness parameters. It is found that the MARS model only performs slightly less accurate than the BPNN model. The computational efficiency of MARS is more superior to the latter. The interpretability of the developed MARS model is demonstrated. In addition, wall deflections computed by this MARS method compare favorably with a number of field and published records.

Table 5.6 Summary of excavation case histories

Case no.	Case name	B (m)	T (m)	H_e (m)	$\frac{c_u}{\sigma_v}$	$\frac{E_{50}}{c_u}$	$\ln\left(\frac{EI}{\gamma_w H_{avg}^4}\right)$	γ (kN/m³)	μ_w	$\delta_{hm\_MARS}^*$ (mm)	$\delta_{h,M}$ (mm)	References
1	Song-san excavation	42	20	9.31	0.25	150	6.24	17.6	1.0	139	105	Ou et al. (2008a, b)
2	TUH in Taiwan	140	40	15.7	0.25	100	7.80	20.0	0.8	47	81	Liao and Hsieh (2002)
3	East Taipei basin	68	23.5	14.1	0.33	100	6.51	16.5	0.8	136	178	Fang (1987)
4	Taiwan Power Company	60	13.5	14.7	0.3	150	6.65	19	0.9	46	63	Moh and Song 2013
5	Shandao Temple	21.5	26.5	18.5	0.3	250	7.82	18.7	0.9	35	37	Hwang et al. (2012)
6	Bugis MRT	21	35	18	0.25	150	8.18	16.5	0.9	125	135	Shirlaw and Wen (1999)
7	Lavender	24	16	15.7	0.25	150	7.96	17.0	0.8	35	32	Lim et al. (2003)
8	Syed Alwi	28	16	7.8	0.25	150	6.43	16.0	0.9	76	50	Lim et al. (2003)
9	MRT line in Singapore	20	20	16	0.25	150	8.11	17.6	0.8	31	39	Goh et al. (2003)
10	Farrer Park	21	22	17.5	0.25	150	7.30	17.3	0.8	76	53	Li (2001)

(continued)

Table 5.6 (continued)

Case no.	Case name	B (m)	T (m)	H_e (m)	$\frac{c_u}{\sigma_v}$	$\frac{E_{50}}{c_u}$	$\ln\left(\frac{EI}{\gamma_w h_{avg}^4}\right)$	γ (kN/m^3)	μ_w	$\delta_{hm\_MARS}^{*}$ (mm)	$\delta_{h,M}$ (mm)	References
11	Rochor Complex	95	24	8.3	0.25	150	4.02	16.0	0.8	248	150	Lim et al. (2003)
12	HDR-4 subway	12.2	22	12.2	0.22	150	6.30	19.1	0.9	103	165	Finno and Harahap (1991)
13	Flagship Wharf	34	8	14	0.2	110	6.80	18.5	0.9	72	110	Konstantakos (1998)
14	Muni Metro Turnback	16	20.5	13.1	0.22	250	7.31	16.5	0.8	59	48	Koutsoftas et al. (2000)
15	Lurie	64	7.4	11.8	0.25	165	5.85	18.9	0.8	61	66	Kung et al. (2007b)
16	Shanghai Bank building	43	19.3	15.2	0.3	190	6.57	18.6	0.9	70	67	Xu et al. (2005)
17	Yanchang Road	18.1	15.5	15.2	0.30	190	5.77	18.0	0.9	45	65	Wang et al. (2005)
18	Pudian Road	20.4	15.5	16.5	0.30	190	6.12	18.0	0.9	34	71	Wang et al. (2005)
19	Kotoku	30	30	17	0.34	100	7.3	16.0	0.8	128	170	Miyoshi (1977)

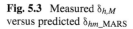**Fig. 5.3** Measured $\delta_{h,M}$
versus predicted δ_{hm_MARS}

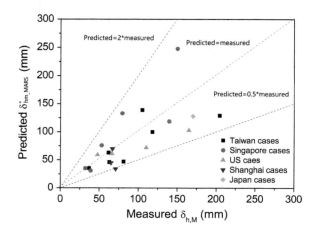

References

Addenbrooke TI, Potts DM, Dabee B (2000) Displacement flexibility number for multiple retaining
    wall design. J Geotech Geoenviron Eng 126(8):718–726
Benz T (2007) Small-strain stiffness of soil and its numerical consequences. PhD thesis, University
    of Stuttgart
Bryson L, Zapata-Medina D (2012) Method for estimating system stiffness for excavation support
    walls. J Geotechnical Geoenviron Eng 138:1104–1115
Burland JB (1989) Small is beautiful-the stiffness of soils at small strains. Can Geotech J
    26(4):499–516
Clough GW, O'Rourke TD (1990) Construction induced movements of in situ walls. In: Design
    and performance of earth retaining structures, ASCE special conference, Ithaca, New York, pp
    439–470
Fang ML (1987) A deep excavation in Taipei Basin. In: Ninth Southeast Asian geotechnical con-
    ference, vol 1, Bangkok, pp 35–42
Finno RJ, Harahap IS (1991) Finite-element analyses of HDR-4 excavation. J Geotech Eng
    117(10):1590–1609
Goh ATC, Wong KS, Teh CI, Wen D (2003) Pile response adjacent to braced excavation. J Geotech
    Geoenviron Eng 129:383–386
Goh ATC, Xuan F, Zhang WG (2013) Reliability assessment of diaphragm wall deflections in soft
    clays. In: Proceedings of foundation engineering in the face of uncertainty, 3–7 March, San Diego,
    GSP No. 229, pp 487–496
Goh ATC, Fan Zhang, Zhang WG, Zhang YM, Hanlong Liu (2017a) A simple estimation model
    for 3D braced excavation wall deflection. Comput Geotech 83:106–113
Goh ATC, Zhang YM, Zhang RH, Zhang WG, Xiao Y (2017b) Evaluating stability of underground
    entry-type excavations using multivariate adaptive regression splines and logistic regression. Tunn
    Undergr Space Technol 70:148–154
Hashash YMA, Whittle AJ (1996) Ground movement prediction for deep excavations in soft clay.
    J Geotech Eng 122(6):474–486
Hsieh PG, Ou CY (2016) Simplified approach to estimate the maximum wall deflection for deep
    excavations with cross walls in clay under the undrained condition. Acta Geotech 11:177–189
Hwang RN, Lee TY, Chou CR, Su TC (2012) Evaluation of performance of diaphragm walls by
    wall deflection paths. J GeoEng 7:1–12

Jardine RJ, Potts DM, Fourie AB, Burland JB (1986) Studies of the influence of non-linear stress-strain characteristics in soil-structure interaction. Géotechnique 36(3):377–396

Konstantakos DC (1998) Measured performance of slurry wall. PhD thesis, Massachusetts Institute of Technology, Cambridge, MA

Koutsoftas DC, Frobenius P, Wu CL, Meyersohn D, Kulesza R (2000) Deformations during cut-and cover construction of MUNI Metro Turnback project. J Geotech Geoenviron Eng 126:344–359

Kung GTC, Hsiao ECL, Juang CH (2007a) Evaluation of a simplified small-strain soil model for analysis of excavation-induced movements. Can Geotech J 44:726–736

Kung GTC, Juang CH, Hsiao ECL, Hashash YMA (2007b) Simplified model for wall deflection and ground-surface settlement caused by braced excavation in clays. J Geotechnical Geoenviron Eng 133(6):731–747

Kung GTC, Ou CY, Juang CH (2009) Modeling small-strain behavior of Taipei clays for finite element analysis of braced excavations. Comput Geotech 36(1):304–319

Li W (2001) Braced excavation in old alluvium in Singapore. Ph.D. thesis, Nanyang Technological University, Singapore

Liao HJ, Hsieh PG (2002) Tied-back excavations in alluvial soil of Taipei. J Geotech Geoenviron Eng 128:435–441

Lim KW, Wong KS, Orihara K, Ng PB (2003) Comparison of results of excavation analysis using WALLUP, SAGE CRISP, and EXCAV97. In: Proceedings of Singapore Underground, pp 83–94

Mana AI, Clough GW (1981) Prediction of movement for braced cuts in clay. J Geotech Geoenviron Eng 107(6):759–777

Miyoshi M (1977) Mechanical behavior of temporary braced wall. In: Proceedings of the 6th international conference on soil mechanics and foundation engineering, vol 2, no 2, Tokyo, pp 655–658

Moh ZC, Song TF (2013) Performance of diaphragm walls in deep foundation excavations. In: First international conferences on case histories in geotechnical engineering 2013, Missouri University of Science and Technology, pp 1335–1343

Moormann C (2004) Analysis of wall and ground movements due to deep excavations in soft soil based on a new worldwide database. Soils Found 44:87–98

Osman AS, Bolton MD (2006) Ground movement predictions for braced excavations in undrained clay. J Geotech Geoenviron Eng 132(4):465–477

Ou CY, Teng FC, Seed RB, Wang IW (2008a) Using buttress walls to reduce excavation-induced movements. Proc ICE-Geotech Eng 161(4):209–222

Ou CY, Teng FC, Wang IW (2008b) Analysis and design of partial ground improvement in deep excavations. Comput Geotech 35(4):576–584

Shirlaw JN, Wen D (1999) Measurements of pore pressure changes and settlements due to a deep excavation in Singapore marine clay. In: Proceedings of Field Measurements in Geomechanics, Balkema, Singapore, pp 241–246

Wang ZW, Ng CWW, Liu GB (2005) Characteristics of wall deflections and ground surface settlements in Shanghai. Can Geotech J 42:1243–1254

Wong KS, Broms BB (1989) Lateral wall deflections of braced excavation in clay. J Geotech Eng 115(6):853–870

Wong I, Poh T, Chuah H (1997) Performance of excavations for depressed expressway in Singapore. J Geotech Geoenviron Eng 123:617–625

Xu ZH, Wang WD, Wang JH, Shen SL (2005) Performance of deep excavation retaining wall in Shanghai soft deposit. Lowland Technol Int 7:31–43

Xuan F (2009) Behavior of diaphragm walls in clays and reliability analysis. M.E. thesis, Nanyang Technological University, Singapore

Zhang WG, Goh ATC, Xuan F (2015a) A simple prediction model for wall deflection caused by braced excavation in clays. Comput Geotech 63:67–72

Zhang WG, Goh ATC, Zhang YM, Chen YM, Xiao Y (2015b) Assessment of soil liquefaction based on capacity energy concept and multivariate adaptive regression splines. Eng Geol 188:29–37

Chapter 6
MARS Use for Inverse Analysis of Soil and Wall Properties in Braced Excavation

As mentioned in Chap. 5, a major concern in deep excavation project in soft clay deposits is the potential for adjacent buildings to be damaged as a result of the associated excessive ground movements. In order to accurately determine the wall deflections using a numerical procedure such as the finite element method, it is critical to use the correct soil parameters such as the stiffness/strength properties. On top of the laboratory and field tests, this can be carried out by performing an inverse analysis using the measured wall deflections. Based on the FE results in Chap. 5, MARS models were developed for inverse parameter identification of the soil relative stiffness ratio and the wall system stiffness, to enable designers to determine the appropriate wall size during the preliminary design phase. Soil relative stiffness ratios and system stiffness values derived via these two different MARS models were found to compare favorably with a number of field and published records.

6.1 Background

For excavations in the ground that comprises of thick soft clays overlying stiff clay, braced walls are usually used to minimize the ground movements. It is common to extend the wall length into the stiff clay layer to prevent basal heave failure and to reduce the movement of the wall toe. To ensure the serviceability limit state is satisfied, a common design criterion is to limit the maximum wall deflection to a fraction of the excavation depth, typically in the range of 0.5–1.5% of the excavation depth H_e. Unnecessarily, severe restrictions may lead to uneconomic design. Therefore, reliable estimates of wall deflections under working conditions are essential.

Numerical tools such as the finite element method are increasingly being used to analyze deep excavation problems. They can provide a better understanding of soil behavior during construction, verify the performance of complex excavations through comparison with field observations, and even predict future responses. Nevertheless, accurate prediction of deformations induced by deep excavations using numerical

© Science Press and Springer Nature Singapore Pte Ltd. 2019
W. Zhang, *MARS Applications in Geotechnical Engineering Systems*,
https://doi.org/10.1007/978-981-13-7422-7_6

approaches is still rather complicated for engineers since apart from modeling the actual construction sequence and wall system parameters, reliable information on the selection of constitutive models and the appropriate soil parameters is also required. Although comprehensive laboratory and field tests can be conducted, there are still some difficulties in the precise determination of some of the soil parameters such as the soil stiffness because of sample disturbance and testing errors. Furthermore, even with well-measured soil parameters, the estimated performance may still deviate from the field observation as a result of the inherent spatial variability and inadequacy of the simulation model itself (Zhao et al. 2015). Therefore, inverse analysis can play a vital role to estimate the relevant soil parameters for more reliable predictions of the expected wall and ground movements that are induced during excavation. Inverse analysis involves utilizing field measurements in order to obtain soil material parameters in contrast to the conventional forward approach. A forward analysis starts with the determination of a constitutive model and its associated parameters derived from laboratory and field testing or empirical relationships. These parameters are then adopted as inputs for numerical analysis to predict stresses, strains, displacements, etc. Previous applications of inverse analyses in geomechanics for soil parameter identification include Gioda (1985), Zentar et al. (2001), Lecampion et al. (2002), Calvello and Finno (2004), Finno and Calvello (2005), Miranda (2007), Levasseur et al. (2008), Rechea et al. (2008), Yan et al. (2009), Papon et al. (2011), Chiu et al. (2012), and Moreira et al. (2013). Optimization algorithms including the gradient-based artificial neural networks and the genetic algorithms are commonly adopted to derive these parameters.

When applying inverse analysis techniques to study the behavior of an actual supported excavation, concerns arise about the proper representation of the real system, as well as the efficiency of the inverse analysis technique and its ability to find a unique set of parameters for a particular geological subsurface. Inverse analyses of supported excavation systems have been carried out by a number of researchers including Ou and Tang (1994), Calvello and Finno (2004), Finno and Calvello (2005), Levasseur et al. (2008), (2010), Hashash et al. (2010), Juang et al. (2013a, b), and Moreira et al. (2013).

Deep excavations in thick deposits of soft clay can cause excessive ground movements and result in damage to adjacent buildings. Numerical analysis using the finite element method to estimate wall deflections for braced excavations can often differ from the values measured in the field. This can be due to many uncertainties with regard to the true soil properties during the preliminary design phase. The use of inverse analysis based on field measurements of wall deflections is therefore a useful technique to infer the correct soil material response, which subsequently can be used to improve the numerical predictions for forward analysis of subsequent excavation stages and for future projects in similar soil conditions.

6.2 The Databases

The database used for the MARS inverse analyses was based on plane strain FE forward analyses of the maximum wall deflection for multistrutted diaphragm walls as described in detail in Zhang et al. (2015a, b) and Chap. 5. However, unlike the data sets in Chap. 5, which is forward analysis results, the data sets used in this chapter are for inverse analysis results; i.e., the calculated/measured wall deflection is used as one of the inputs to back-calculate/calibrate the key soil/wall parameters. A total of 1032 cases were considered, based on parameter combinations of the seven design variables. Some typical data sets including the training and testing patterns are shown in Tables 6.1 and 6.2.

6.3 The Developed MARS E_{50}/c_u Model and Parametric Sensitivity Analysis

MARS model has been developed for inverse analysis to estimate the soil relative stiffness ratio E_{50}/c_u as a function of seven input parameters: γ, B, H_e, c_u/σ'_v, δ_h^*, $\ln(S)$, and T. Herein, the measured wall deflection $\delta_{hm\_calculated}$ is used as an input parameter to back-calculate E_{50}/c_u.

The data set was separated randomly into a training set of 775 patterns and a testing set of 257 patterns. The MARS model with the highest R^2 value and less BFs for the testing data set is considered to be optimal. The optimal MARS model adopted 28 BFs of linear spline functions. The observed versus predicted 1:1 plots of E_{50}/c_u is shown in Fig. 6.1. The developed model predicts slightly higher estimations for low E_{50}/c_u values and slightly lower estimations for higher values of E_{50}/c_u. Figure 6.2 presents the histogram plots of the relative errors e (defined as the ratio of the difference between the MARS predicted and the target E_{50}/c_u to the target value, in percentage). It is obvious that most of the MARS estimations of the data patterns fell within $\pm 20\%$ of the target values. In addition, it should be noted that MARS predictions still have a large range of variations with respect to the same group of targets. On the other hand, it can be acceptable since the target outputs are of category nature (types), while the estimations based on the developed model are numerical values. Some typical training and testing data sets together with the MARS predictions are listed in Tables 6.1 and 6.2, respectively.

Table 6.3 displays the ANOVA decomposition of the developed MARS model. The first column lists the ANOVA function number. The second column gives an indication of the importance of the corresponding ANOVA function, by listing the GCV score for a model with all BFs corresponding to that particular ANOVA function removed. The third column provides the standard deviation of this function. The fourth column gives the number of BFs comprising of the ANOVA function. The last column gives the particular input variables associated with the ANOVA function.

Table 6.1 Some typical training data for MARS E_{50}/c_u identification model

B (m)	T (m)	H_e (m)	c_u/σ'_v	δ^*_h (mm)	$\ln(S)$	γ (kN/m^3)	Target E_{50}/c_u	MARS predicted E_{50}/c_u
30	30	11	0.34	26	8.846	19.0	400	390
30	30	11	0.34	35	7.313	19.0	400	405
40	25	14	0.34	48	8.176	19.0	300	281
30	25	20	0.34	70	7.313	17.0	300	280
40	25	20	0.25	76	8.176	19.0	300	274
30	30	20	0.29	77	8.176	19.0	300	286
30	30	20	0.34	77	7.313	19.0	300	300
40	25	20	0.25	98	7.313	19.0	300	281
40	30	20	0.29	98	8.176	17.0	300	283
30	30	11	0.29	54	7.313	19.0	300	303
40	25	14	0.29	69	8.176	19.0	200	211
40	35	20	0.34	100	8.176	19.0	200	201
50	25	20	0.34	101	7.313	17.0	200	232
30	30	20	0.34	101	8.176	15.0	200	229
40	30	17	0.34	92	8.176	15.0	300	287
40	30	20	0.21	115	8.846	19.0	200	215
50	25	14	0.29	81	8.176	17.0	200	210
40	35	14	0.34	81	6.097	19.0	300	313
40	30	17	0.34	99	8.176	19.0	100	102
40	30	17	0.21	107	8.846	19.0	200	225
40	30	17	0.29	108	8.176	15.0	300	291
30	30	20	0.25	134	8.176	15.0	200	224
40	30	11	0.25	74	8.176	17.0	300	290
40	30	20	0.34	135	7.313	15.0	300	260
30	30	20	0.29	135	7.313	17.0	200	208
30	30	20	0.34	298	6.097	15.0	100	101
40	30	20	0.25	293	6.097	17.0	200	200
50	35	14	0.29	204	6.097	17.0	200	219
40	35	11	0.25	159	7.313	19.0	100	87
50	25	11	0.25	150	7.313	17.0	100	95

Table 6.2 Some typical testing data for MARS E_{50}/c_u identification model

B (m)	T (m)	H_e (m)	c_u/σ'_v	δ_h^* (mm)	$\ln(S)$	γ (kN/m^3)	Target E_{50}/c_u	MARS predicted E_{50}/c_u
30	30	11	0.29	37	8.176	19.0	400	395
40	30	11	0.34	42	8.176	19.0	300	326
40	25	17	0.34	65	8.176	19.0	200	207
30	30	14	0.34	55	7.313	19.0	300	290
60	30	14	0.29	76	8.846	17.0	200	246
40	30	20	0.25	115	8.176	17.0	300	269
30	30	17	0.25	98	8.176	19.0	200	230
40	25	20	0.29	119	6.097	19.0	300	276
30	30	17	0.34	112	7.313	15.0	300	263
40	30	20	0.25	132	8.176	15.0	300	275
40	25	17	0.34	114	7.313	19.0	100	114
30	30	17	0.29	125	8.176	17.0	100	102
30	30	17	0.34	167	6.097	19.0	100	87
40	30	14	0.29	138	6.097	19.0	200	211
30	25	11	0.29	110	6.097	17.0	200	207
50	25	11	0.29	74	7.33	17.0	300	293

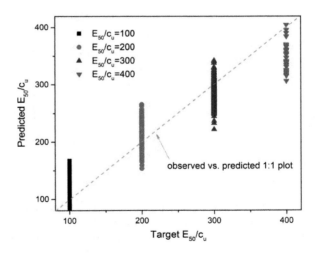

Fig. 6.1 Observed versus predicted 1:1 plots of E_{50}/c_u

Fig. 6.2 Histogram of relative errors for the MARS E_{50}/c_u model

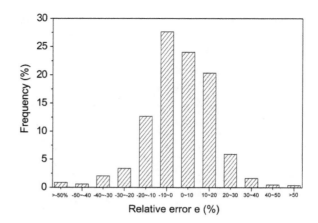

Table 6.3 ANOVA decomposition for MARS E_{50}/c_u model

Function	GCV	STD	#Basis	Variable(s)
1	1711.6	23.61	2	B
2	4915.2	58.34	2	T
3	5724.9	54.83	2	H_e
4	9107.5	54.66	2	c_u/σ'_v
5	25614.0	134.44	2	δ_h^*
6	12670.7	98.30	2	$\ln(S)$
7	2584.5	37.73	2	γ
8	1290.3	19.34	2	$T \& H_e$
9	1027.5	12.38	2	$T \& c_u/\sigma'_v$
10	944.8	8.98	2	$T \& \gamma$
11	955.4	9.62	2	$H_e \& \ln(S)$
12	1029.3	12.46	2	$c_u/\sigma'_v \& \ln(S)$
13	1289.3	17.04	1	$\delta_h^* \& \ln(S)$
14	983.2	10.42	1	$\delta_h^* \& \gamma$
15	1355.8	20.12	2	$\ln(S) \& \gamma$

Figure 6.3 gives the plot of the relative importance of the input variables, which is evaluated by the increase in the GCV value caused by removing the considered variables from the developed MARS model. The results indicate that the three most important variables influencing the determination of E_{50}/c_u are the calculated wall deflection δ_h^*, the system stiffness in a logarithmic scale $\ln(S)$, and the soil relative shear strength ratio c_u/σ'_v.

Table 6.4 lists the BFs and their corresponding equations for the developed MARS model. It is observed from Table 6.4 that interactions have occurred between BFs since exactly half of the basis functions are products of two spline functions (from

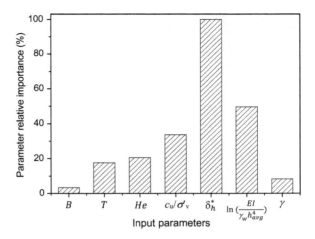

Fig. 6.3 Relative importance of the input variables selected in the MARS E_{50}/c_u model

BF15 to BF28). The presence of interactions suggests that these two models are not simply additive and that interactions play an important role in building an accurate model for soil parameter identification. The equation of this optimal MARS model is given by

$$
\begin{aligned}
E_{50}/c_u = {}& 290 - 1.97 \times BF1 + 3.45 \times BF2 - 133.26 \times BF3 + 106.75 \times BF4 \\
& + 8.61 \times BF5 - 19.86 \times BF6 - 1377 \times BF7 + 1575 \times BF8 - 17.64 \\
& \times BF9 + 37.12 \times BF10 + 15.80 \times BF11 - 23.47 \times BF12 + 1.2 \\
& \times BF13 - 4.67 \times BF14 + 0.67 \times BF15 - 1.69 \times BF16 + 2.16 \\
& \times BF17 + 2.75 \times BF18 - 19.37 \times BF19 + 31.61 \times BF20 - 372 \\
& \times BF21 + 967 \times BF22 - 53.95 \times BF23 + 85 \times BF24 - 2.42 \\
& \times BF25 + 2.8 \times BF26 + 3.23 \times BF27 - 8.9 \times BF28
\end{aligned}
\tag{6.1}
$$

6.4 Excavation History Validations on E_{50}/c_u Model

To validate this proposed MARS model for soil parameter identification, a total of 12 well-documented excavation case histories from various countries as listed in Table 6.5 were analyzed. To visualize the quality of prediction, predicted stiffness ratio E_{50}/c_u values by MARS model are compared with the target values for the 12 cases listed in Table 6.5 and shown in Fig. 6.4. Figure 6.4 also plots the range of E_{50}/c_u by varying δ_h^* by +15% and −15%. Table 6.5 and Fig. 6.4 indicate that the developed MARS model is able to predict reasonably well the soil stiffness ratios for the case histories considered, even with considerable variability in the wall deflection measurements.

Table 6.4 Basis functions and corresponding equations of MARS model for E_{50}/c_u

BF	Equation	BF	Equation
BF1	$\max(0, \delta_h^* - 132)$	BF15	BF2 × $\max(0, \ln(S) - 7.313)$
BF2	$\max(0, 132 - \delta_h^*)$	BF16	BF6 × $\max(0, T - 30)$
BF3	$\max(0, \ln(S) - 7.313)$	BF17	BF6 × $\max(0, 30 - T)$
BF4	$\max(0, 7.313 - \ln(S))$	BF18	BF9 × $\max(0, 54 - \delta_h^*)$
BF5	$\max(0, H_e - 17)$	BF19	BF4 × $\max(0, \gamma - 17)$
BF6	$\max(0, 17 - H_e)$	BF20	BF4 × $\max(0, 17 - \gamma)$
BF7	$\max(0, c_u/\sigma_v' - 0.25)$	BF21	BF4 × $\max(0, c_u/\sigma_v' - 0.29)$
BF8	$\max(0, 0.25 - c_u/\sigma_v')$	BF22	BF4 × $\max(0, 0.29 - c_u/\sigma_v')$
BF9	$\max(0, \gamma - 17)$	BF23	BF7 × $\max(0, T - 30)$
BF10	$\max(0, 17 - \gamma)$	BF24	BF7 × $\max(0, 30 - T)$
BF11	$\max(0, T - 30)$	BF25	BF11 × $\max(0, \gamma - 17)$
BF12	$\mathrm{Max}(0, 30 - T)$	BF26	BF11 × $\max(0, 17 - \gamma)$
BF13	$\max(0, B - 40)$	BF27	BF4 × $\max(0, H_e - 14)$
BF14	$\max(0, 40 - B)$	BF28	BF4 × $\max(0, 14 - H_e)$

Fig. 6.4 Target and predicted E_{50}/c_u values by developed MARS model

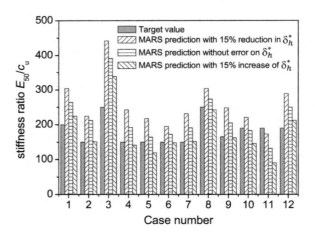

Table 6.5 Summary of excavation case histories validating MARS E_{50}/c_u model

Case no.	Case name	B (m)	T (m)	H_e (m)	$\frac{c_u}{\sigma_v}$	$\delta_{hm,\text{measured}}$ (mm)	$\ln(S)$	γ (kN/m³)	μ_w^a	Target $\frac{E_{50}}{c_u}$ [b]	MARS predictions	References
1	Taiwan Power Company	60	13.5	14.7	0.30	63	6.65	19.0	0.9	150	212	Moh and Song (2013)
2	Shandao Temple	21.5	26.5	18.5	0.30	36.7	7.82	18.7	0.8	250	391	Fang (1987)
3	Xinyi Planning Zone	41	27	14.45	0.34	78	7.02	18.7	0.8	150	192	Fang et al. (2004)
4	Bugis MRT	21	35	18	0.25	135	8.18	16.5	0.9	150	164	Li (2001)
5	Lavender	24	16	15.7	0.25	32	7.96	17.0	0.8	150	172	Lim et al. (2003)
6	MRT line in Singapore	20	20	16	0.25	38.6	8.11	17.6	0.8	150	191	Goh et al. (2003)
7	Muni Metro Turnback	16	20.5	13.1	0.22	48	7.31	16.5	0.8	250	273	Koutsoftas et al. (2000)
8	Lurie	64	7.4	11.8	0.25	66	5.85	18.9	0.8	165	205	Kung et al. (2007)
9	Yanchang Road	18.1	15.5	15.2	0.30	65	5.77	18.0	0.9	190	183	Wang et al. (2005)
10	Pudian Road	20.4	15.5	16.5	0.30	71	6.12	18.0	0.9	190	132	Wang et al. (2005)

(continued)

Table 6.5 (continued)

Case no.	Case name	B (m)	T (m)	H_e (m)	$\frac{c_u}{\sigma_v}$	$\delta_{hm,\text{measured}}$ (mm)	$\ln(S)$	γ (kN/m^3)	μ_w^a	Target $\frac{E_{50}}{c_u}$ [b]	MARS predictions	References
11	Shanghai Bank building	43	19.3	15.2	0.30	67.4	6.57	18.6	0.9	190	250	Xu et al. (2005)

[a] μ_w, modification factor, which considers the change of groundwater level during excavation
[b] Herein, target means that the values are from either laboratory/field tests and reported by the various authors or back analysis

Fig. 6.5 Distribution of the predicted E_{50}/c_u values based on Eq. (6.1) for Bugis MRT

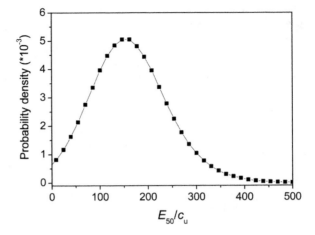

Based on Eq. (6.1), Fig. 6.5 gives the plot of the probability density of E_{50}/c_u for a typical case (Bugis MRT station) derived from Monte Carlo simulation with 1,000,000 iterations assuming that the coefficient of variation COV of both $\delta_{hm\_measured}$ and c_u/σ'_v is 0.15. Both $\delta_{hm\_measured}$ and c_u/σ'_v follow the normal distribution, the mean values of which are 150 and 0.25, respectively. The other five input variables are deterministic, and the values can be found in Table 6.5. As can be seen from Fig. 6.5, the variation of E_{50}/c_u follows the normal distribution. The most probable E_{50}/c_u value is 147, very close to the target value of 150.

6.5 The Developed MARS ln(S) Model and Parametric Sensitivity Analysis

A second MARS model was also developed for inverse parameter estimation of the wall system stiffness. This model will assist engineers to determine the appropriate wall size during the preliminary design phase. The same patterns as used for E_{50}/c_u model are adopted for training and testing of the MARS ln(S) model, respectively. The optimal MARS model adopted 28 BFs of linear spline functions. For comparison, Fig. 6.6 plots the R^2 values of the testing data sets for the MARS ln(S) model with different BFs (from 14 to 48). The observed versus predicted 1:1 plots of ln(S) is shown in Fig. 6.7. Figure 6.8 presents the histogram plots of the relative errors. It is obvious that most of the MARS estimations of the data patterns fell within ±10% of the target values. Some typical training and testing data sets together with the MARS predictions are listed in Tables 6.6 and 6.7, respectively.

Table 6.8 lists the BFs and their corresponding equations for the developed MARS ln(S) model. It is observed from Table 6.8 that interactions have occurred between BFs since exactly 12 out of the 28 BFs are products of two spline functions. The

Fig. 6.6 R^2 for a different
number of BFs for $\ln(S)$
model

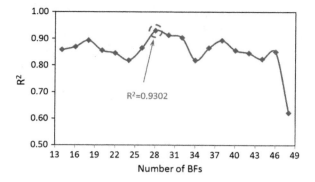

Fig. 6.7 Observed versus
predicted 1:1 plots of $\ln(S)$

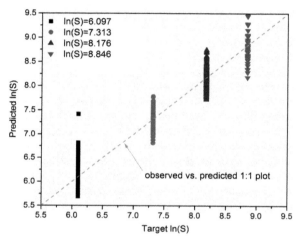

Fig. 6.8 Histogram of
relative errors for MARS
$\ln(S)$ model

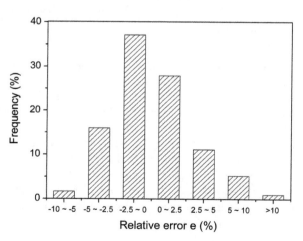

Table 6.6 Some typical training data for MARS $\ln(S)$ model

B (m)	T (m)	H_e (m)	c_u/σ'_v	E_{50}/c_u	δ_h^* (mm)	γ (kN/m^3)	Target $\ln(S)$	MARS predicted $\ln(S)$
40	25	20	0.34	100	95	19.0	8.176	8.319
30	25	20	0.34	100	110	17.0	7.313	7.559
30	30	20	0.34	100	115	17.0	8.176	8.120
30	30	14	0.34	100	98	17.0	8.176	8.169
40	25	20	0.34	200	68	19.0	8.176	8.109
30	30	17	0.34	200	58	19.0	8.846	8.721
30	30	20	0.25	200	86	19.0	8.846	9.153
40	30	20	0.29	200	87	19.0	8.846	8.849
40	30	20	0.21	200	115	19.0	8.846	9.029
50	25	14	0.29	200	81	17.0	8.176	8.251
40	25	20	0.34	300	57	19.0	8.176	8.092
30	30	20	0.34	300	63	19.0	8.176	8.001
30	25	20	0.34	300	70	17.0	7.313	7.499
30	30	20	0.29	300	77	19.0	8.176	8.048
40	25	17	0.25	300	72	19.0	8.176	8.235
30	30	17	0.25	300	83	19.0	8.176	8.145
30	30	11	0.29	300	54	19.0	7.313	7.315
40	25	17	0.25	300	142	19.0	6.097	5.856
40	30	11	0.21	300	93	17.0	8.176	8.030
40	30	17	0.25	300	181	19	6.097	6.142
30	30	14	0.34	300	151	15.0	6.097	5.951
40	30	20	0.25	300	202	19.0	6.097	6.044
30	30	11	0.34	400	26	19.0	8.846	8.699
30	30	17	0.34	400	41	19.0	8.846	8.810
40	30	20	0.34	400	75	19.0	7.313	7.332
40	30	17	0.34	400	68	19.0	7.313	7.351
30	30	11	0.29	400	45	19.0	7.313	7.260
30	30	17	0.29	400	61	19.0	8.176	8.070
40	35	17	0.25	400	123	19.0	7.313	7.399
40	30	14	0.25	400	90	19.0	7.313	7.292

Table 6.7 Some typical testing data for MARS ln(S) model

B (m)	T (m)	H_e (m)	c_u/σ'_v	E_{50}/c_u	δ^*_h (mm)	γ (kN/m³)	Target ln(S)	MARS predicted ln(S)
30	30	20	0.34	100	106	19.0	8.176	8.151
30	30	20	0.29	100	122	19.0	8.176	8.310
40	25	17	0.34	100	114	19.0	7.313	7.012
50	25	20	0.34	100	136	17.0	7.313	7.357
30	30	20	0.29	200	75	19.0	8.846	8.904
30	25	17	0.34	200	78	17.0	7.313	7.281
50	25	17	0.29	100	151	17.0	7.313	7.381
50	25	20	0.29	200	116	17.0	7.313	7.442
40	30	11	0.34	200	64	19.0	7.313	7.441
30	25	17	0.21	200	126	17.0	7.313	7.442
30	25	20	0.29	200	150	17.0	6.097	5.955
40	35	17	0.34	200	128	17.0	7.313	7.337
40	25	17	0.29	300	63	19.0	8.176	8.004
30	30	14	0.34	300	54	19.0	8.176	8.219
40	25	20	0.34	300	94	19.0	6.097	6.075
40	35	20	0.25	400	147	19.0	7.313	7.297

presence of interactions suggests that these two models are not simply additive and that interactions play an important role in building MARS ln(S) model. The equation of this optimal MARS model is given by

$$
\begin{aligned}
\ln(S) = {} & 7.055 + 0.038 \times \text{BF1} - 16.977 \times \text{BF2} + 17.522 \times \text{BF3} - 0.004 \\
& \times \text{BF4} + 0.007 \times \text{BF5} - 0.0007 \times \text{BF6} - 0.318 \times \text{BF7} + 0.439 \\
& \times \text{BF8} - 0.0004 \times \text{BF9} + 0.024 \times \text{BF10} - 0.046 \times \text{BF11} \\
& + 0.149 \times \text{BF12} - 0.225 \times \text{BF13} - 0.007 \times \text{BF14} + 0.03 \\
& \times \text{BF15} - 0.0002 \times \text{BF16} + 0.0003 \times \text{BF17} + 2 \times 10^{-5} \\
& \times \text{BF18} + 0.1 \times \text{BF19} - 0.152 \times \text{BF20} - 0.013 \times \text{BF21} + 0.015 \\
& \times \text{BF22} + 0.010 \times \text{BF23} + 0.040 \times \text{BF24} + 0.259 \\
& \times \text{BF25} + 7 \times 10^{-5} \times \text{BF26} + 0.0007 \times \text{BF27} - 0.0008 \times \text{BF28} \quad (6.2)
\end{aligned}
$$

Table 6.8 Basis functions and corresponding equations of MARS $\ln(S)$ model

BF	Equation	BF	Equation
BF1	$\max(0, 46 - \delta_h^*)$	BF15	$\max(0, 118 - \delta_h^*)$
BF2	$\max(0, c_u/\sigma'_v - 0.25)$	BF16	BF13 $\times \max(0, E_{50}/c_u - 200)$
BF3	$\max(0, 0.25 - c_u/\sigma'_v)$	BF17	BF13 $\times \max(0, 200 - E_{50}/c_u)$
BF4	$\max(0, E_{50}/c_u - 300)$	BF18	$\max(0, \delta_h^* - 46) \times \max(0, E_{50}/c_u - 200)$
BF5	$\max(0, 300 - E_{50}/c_u)$	BF19	$\max(0, T - 30)$
BF6	$\max(0, \delta_h^* - 46) \times \max(0, H_e - 14)$	BF20	$\max(0, 30 - T)$
BF7	$\max(0, \gamma - 17)$	BF21	BF13 $\times \max(0, T - 30)$
BF8	$\max(0, 17 - \gamma)$	BF22	BF13 $\times \max(0, 30 - T)$
BF9	$\max(0, \delta_h^* - 46) \times \max(0, 30 - T)$	BF23	$\max(0, 190 - \delta_h^*)$
BF10	$\max(0, B - 40)$	BF24	BF2 $\times \max(0, \delta_h^* - 74)$
BF11	$\max(0, 40 - B)$	BF25	BF2 $\times \max(0, 74 - \delta_h^*)$
BF12	$\max(0, H_e - 17)$	BF26	BF23 $\times \max(0, 200 - E_{50}/c_u)$
BF13	$\max(0, 17 - H_e)$	BF27	BF5 $\times \max(0, \gamma - 17)$
BF14	$\max(0, \delta_h^* - 118)$	BF28	BF5 $\times \max(0, 17 - \gamma)$

6.6   Excavation History Validations on $\ln(S)$ Model

The same excavation case histories as used for validating the E_{50}/c_u model are adopted for verifying the developed MARS $\ln(S)$ model. The MARS predicted $\ln(S)$ values are compared with the target ones for cases listed in Table 6.9 and shown in Fig. 6.9. Table 6.9 and Fig. 6.9 indicate that the developed MARS model is able to provide reasonable estimates of the wall stiffness for the case histories considered.

6.7   Summary

This chapter presents two MARS models developed for identifying soil parameters and configuring system stiffness in braced excavations based on field observation

Table 6.9 Summary of excavation case histories validating MARS $\ln(S)$ model

Case no.	Case name	B (m)	T (m)	He (m)	$\frac{c_u}{\sigma_v}$	$\frac{E_{50}}{c_u}$	$\delta_{h,h}$ (mm)	γ (kN/m3)	μ_w	Target $\ln(S)$	MARS predicted $\ln(S)$	References
1	Taiwan Power Company	60	13.5	14.7	0.30	150	63	19.0	0.9	6.65	7.786	Moh and Song (2013)
2	Shandao Temple	21.5	26.5	18.5	0.30	250	36.7	18.7	0.8	7.82	8.761	Fang (1987)
3	Xinyi Planning Zone	41	27	14.45	0.34	150	78	18.7	0.8	7.02	7.201	Fang et al. (2004)
4	Bugis MRT	21	35	18	0.25	150	135	16.5	0.9	8.18	8.054	Li (2002)
5	Lavender	24	16	15.7	0.25	150	32	17.0	0.8	7.96	9.729	Lim et al. (2003)
6	MRT line in Singapore	20	20	16	0.25	150	38.6	17.6	0.8	8.11	9.343	Goh et al. (2003)
7	Muni Metro Turnback	16	20.5	13.1	0.22	250	48	16.5	0.8	7.31	8.253	Koutsoftas et al. (2000)
8	Lurie	64	7.4	11.8	0.25	165	66	18.9	0.8	5.85	7.440	Kung et al. (2007)
9	Yanchang Road	18.1	15.5	15.2	0.30	190	65	18.0	0.9	5.77	5.891	Wang et al. (2005)
10	Pudian Road	20.4	15.5	16.5	0.30	190	71	18.0	0.9	6.12	5.575	Wang et al. (2005)
11	Shanghai Bank building	43	19.3	15.2	0.30	190	67.4	18.6	0.9	6.57	7.218	Xu et al. (2005)

Fig. 6.9 Target and predicted ln(S) values by developed MARS model

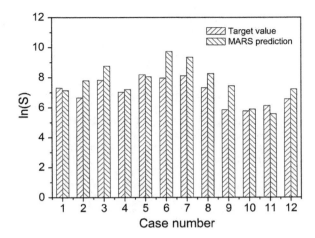

of maximum wall deflections. The MARS E_{50}/c_u model can relate the soil relative stiffness ratio to influencing parameters including the excavation geometries, soil shear strength ratio, unit weight, and the wall stiffness. The developed MARS ln(S) model provides estimates of the suitable wall stiffness based on a user-defined allowable wall deflection and thus saves design time by eliminating unsuitable wall configurations early in the design process. Well-documented case histories from various countries validating the reliability of the proposed MARS models are given for a wide variety of wall and soil conditions.

References

Calvello M, Finno RJ (2004) Selecting parameters to optimize in model calibration by inverse analysis. Comput Geotech 31(5):411–425

Chiu CF, Yan WM, Yuen KV (2012) Estimation of water retention curve of granular soils from particle size distribution—a Bayesian probabilistic approach. Can Geotech J 49(9):1024–1035

Fang ML (1987) A deep excavation in Taipei Basin. In: Ninth Southeast Asian Geotechnical Conference 1987, Bangkok, 1:35–42

Fang TC, Tsai YY, Su TC, Tsung P, Seeley P (2004) A case study on time-dependent displacement of diaphragm wall induced by creep of soft clay. In: Proceedings of 5th cross-strait Geotechnics Seminars, 9–11 Nov. 2004; Taipei, pp 283–290

Finno RJ, Calvello M (2005) Supported excavations: observational method and inverse modeling. J Geotechn Geoenvironmental Eng 131(7):826–836

Gioda G (1985) Some remarks on back analysis and characterization problems in geomechanics. In: Proceedings of 5th international conference on numerical methods in geomechanics 1985; Nagoya, Japan. Balkema, Rotterdam, pp 47–61

Goh ATC, Wong KS, Teh CI, Wen D (2003) Pile response adjacent to braced excavation. J Geotechn Geoenvironmental Eng 129:383–386

Hashash Y, Levasseur S, Osouli A, Finno R, Malecot Y (2010) Comparison of two inverse analysis techniques for learning deep excavation response. Comput Geotech 37:323–333

Juang CH, Ching J, Wang L, Khoshnevisan S, Ku CS (2013a) Simplified procedure for estimation of liquefaction-induced settlement and site-specific probabilistic settlement exceedance curve using cone penetration test (CPT). Can Geotech J 50(10):1055–1066

Juang CH, Luo Z, Atamturktur S, Huang H (2013b) Bayesian updating of soil parameters for braced excavations using field observations. J Geotechn Geoenvironmental Eng 139:395–406

Koutsoftas DC, Frobenius P, Wu CL, Meyersohn D, Kulesza R (2000) Deformations during cut- and cover construction of MUNI Metro Turnback project. J Geotechn Geoenvironmental Eng 126:344–359

Kung GTC, Hsiao ECL, Juang CH (2007) Evaluation of a simplified small-strain soil model for analysis of excavation-induced movements. Can Geotech J 44:726–736

Lecampion B, Constantinescu A, Nguyen Minh D (2002) Parameter identification for lined tunnels in viscoplastic medium. Int J Numer Anal Meth Geomech 26:1191–1211

Levasseur S, Malécot Y, Boulon M, Flavigny E (2008) Soil parameter identification using a genetic algorithm. Int J Numer Anal Meth Geomech 32:189–213

Levasseur S, Malécot Y, Boulon M, Lavigny E (2010) Statistical inverse analysis based on genetic algorithm and principal component analysis: applications to excavation problems and pressureme- ter tests. Int J Numer Anal Meth Geomech 34:471–491

Lim KW, Wong KS, Orihara K, Ng PB (2003) Comparison of results of excavation analysis using WALLUP, SAGE CRISP, and EXCAV97. In: Proceedings of Singapore Underground 2003, pp 83–94

Miranda T (2007) Geomechanical parameters evaluation in underground structures. Artificial intel- ligence, Bayesian probabilities and inverse methods, Ph.D.'s thesis 2007. University of Minho, Guimarães, Portugal

Moh ZC, Song TF (2013) Performance of diaphragm walls in deep foundation excavations. In: First international conferences on case histories in geotechnical engineering 2013, Missouri University of Science and Technology, pp 1335–1343

Moreira N, Miranda T, Pinheiro M, Fernandes P, Dias D, Costa L, Sena-Cruz J (2013) Back anal- ysis of geomechanical parameters in underground works using an Evolution Strategy algorithm. Tunneling Underground Space Technol 33:143–158

Ou CY, Tang Y (1994) Soil parameter determination for deep excavation analysis by optimization. J Chinese Inst Eng 17(5):671–688

Papon A, Riou Y, Dano C, Hicher PY (2011) Single and multi-objective genetic algorithm opti- mization for identifying soil parameters. Int J Numer Anal Meth Geomech 36:597–618

Rechea C, Levasseur S, Finno R (2008) Inverse analysis techniques for parameter identification in simulation of excavation support systems. Comput Geotech 35(3):331–345

Wang ZW, Ng CWW, Liu GB (2005) Characteristics of wall deflections and ground surface settle- ments in Shanghai. Can Geotech J 42:1243–1254

Xu ZH, Wang WD, Wang JH, Shen SL (2005) Performance of deep excavation retaining wall in Shanghai soft deposit. Lowland Technol Int 7:31–43

Yan WM, Yuen KV, Yoon GL (2009) Bayesian probabilistic approach for the correlations of com- pressibility index for marine clays. J Geotech Geoenvironmental Eng 135(12):1932–1940

Zentar R, Hicher P, Moulin G (2001) Identification of soil parameters by inverse analysis. Comput Geotech 28:129–144

Zhang WG, Goh ATC, Xuan F (2015a) A simple prediction model for wall deflection caused by braced excavation in clays. Comput Geotech 63:67–72

Zhang WG, Goh ATC, Zhang YM, Chen YM, Xiao Y (2015b) Assessment of soil liquefaction based on capacity energy concept and multivariate adaptive regression splines. Eng Geol 188:29–37

Zhao BD, Zhang LL, Jeng DS, Wang JH, Chen JJ (2015) Inverse analysis of deep excavation using differential evolution algorithm. Int J Numer Anal Meth Geomech 39:115–134

Chapter 7
MARS Use for Estimations of Lateral Wall Deflection Profiles

This chapter presented MARS approach for estimating wall deflection profile caused by deep braced excavations, based on an expanded database including a total of 30 case histories for braced excavation in stiff, medium, and soft clays. Seven input variables, including wall length, excavation depth, excavation length, system stiffness, average unit weight, undrained shear strength of the soil, and the depth below the ground surface, are adopted as inputs to the MARS deflection profile model. Comparison with two more excavation case histories indicates that the developed MARS model can give an accurate graphical representation of the wall deflection profile. It is capable of not only predicting the value of maximum wall deflection but also estimating the possible depth at which maximum lateral deformation occurs.

7.1 Background

With the rapid growth in urban development, more and more excavation projects for high-rise buildings and subway lines are being executed and scheduled. In the design of a braced excavation, it is essential to consider not only the stability issue such as basal heave during construction, but also the potential serviceability problem of adjacent buildings caused by excessive wall deflections and ground movements. In order to ensure the stability of the excavation and reduce the effect on the neighboring buildings and underground utilities caused by excavation, continuous wall structures are often used. In these cases, the use of a multistrutted structural system is generally required in order to reduce ground movements and also to achieve relatively high benefits. A reasonable estimation of the lateral wall deflection profiles caused by braced excavations and the corresponding maximum value is desirable for safe and economical design purpose.

There are two common techniques for estimating the horizontal wall displacements and ground settlements using either interpolation from a published database of different areas of the world or numerical analysis using either finite element methods or finite difference methods. Since soil is a complicated material that always

© Science Press and Springer Nature Singapore Pte Ltd. 2019
W. Zhang, *MARS Applications in Geotechnical Engineering Systems*,
https://doi.org/10.1007/978-981-13-7422-7_7

displays nonlinear and inelastic behavior, robust predictions of ground movements are difficult. Even though various aspects of soil are incorporated into many numerical models, many of these models are generally complex and the adopted parameters do not have a clear physical meaning, and even the determination of them requires special kind of testing technique and laboratory skills. In addition, these numerical models require a huge amount of computational resources. Actually, the performance of deep excavations depends on a large number of geotechnical and geometrical influential parameters that are interdependent to different degrees. Therefore, practicing engineers are apt to avoid using numerical simulations and tend to use design charts which directly relate wall deflections to soil properties. It is a proven approach to identify the significant parameters affecting the deformation behavior of deep excavations through the empirical analysis of the measured displacements in a large number of case histories.

Currently, the most commonly used design chart for predicting lateral wall deflections in deep excavations in clay is the one proposed by Clough et al. (1989). It allows the estimation of lateral movements in terms of system stiffness which is principally represented by three factors including the bending stiffness of the retaining wall and supports, the configuration, location and distance of the strut, the embedment length of the retaining wall, and the factor of safety against basal heave. Long (2001) and Moormann (2004) have assessed the validity and applicability of the Clough design chart using databases of more than 296 and 530 case histories, respectively, plotting maximum lateral deformation, normalized with respect to the excavation height, versus system stiffness, and compared the result with the curves proposed by Clough et al. (1989). For different factors of safety against basal heave, Long (2001) differentiated the data by a low and high factor of safety while Moormann (2004) differentiated it by soft and stiff ground.

Several other empirical and semiempirical methods commonly adopted for estimating maximum wall deflections can be referred to in Mana and Clough (1981), Wong and Broms (1989), Clough and O'Rourke (1990), Hashash and Whittle (1996), Addenbrooke et al. (2000), Kung et al. (2007), Bryson and Zapata (2012), and Zhang et al. (2015a, b). However, most of these methods aim at solving for the lateral maximum wall deflection, without any attention on the lateral wall deflection profiles, which is thought to be of the same significance with the maximum value. This chapter utilized the MARS method to derive the estimation models for the lateral wall deflection profiles and also the maximum wall deflections induced by excavations in clays, based on an expanded database with different clay types, excavation geometries, soil, and support system parameters. Validations of the proposed MARS model were also carried out through comparisons with two more well-documented excavation case histories.

7.2   The Databases

The databases developed by Long (2001) and Moormann (2004) are mostly used. Based on the results of his empirical study, Moormann (2004) concluded that the data for deep excavations in soft clays scatter in a wide range and there is no clear dependency of the system stiffness factor proposed by Clough et al. (1989) on the lateral deflections. For stiff clays, the results are similar to those presented by Long (2001) where the displacements are not influenced by a factor of safety against basal heave and their dependency on the system stiffness is not observed. Moormann (2004) regarded the lack of dependency of lateral wall deflections on system stiffness to factors including the soil conditions at the embedment portion of the wall, the groundwater conditions, the surrounding buildings or geometrical irregularities, workmanship, excavation sequence, and pre-load of struts. However, the quantification of these factors mentioned above is difficult since they are not reported and documented in detail in most cases. Due to this and also in view of the lack of information in the case histories presented by Long (2001) and Moormann (2004), a detailed database is essentially needed for investigating the aforementioned factors that may influence the lateral wall deflections in a deep excavation in clays.

Zapata (2007) compiled the expended database with case histories distinguished by soil type defined by the undrained shear strength found at the dredge level of the excavation. Table 7.1 lists the locations and the references of these case histories. Note that ten case histories are presented for each soil type, giving a total of 30 case histories. For detailed information about the subsurface soil conditions, geometry characteristics, excavation support system details, raw inclinometer data, and maximum ground movements for each case history, Zapata (2007) is referred.

Table 7.2 summarizes the geometric (H_w, H_e, and B), soil ($\gamma_{s,avg}$ and s_u), and support system $\left(EI/\gamma_w S_v^4\right)$ parameters for the case histories on stiff, medium, and soft clay, respectively. In addition, the maximum horizontal wall movements recorded at the end of excavation are also included.

Figure 7.1 shows typically the subsurface soil conditions, geometry characteristics, excavation support system details, raw inclinometer data, and maximum ground movements for case history no. 1: Lion Yard Development in Cambridge, UK (Ng 1992).

Based on each case history listed in Table 7.2 and the corresponding wall deflection profiles against depth below the ground surface, the MARS model is developed. A total of seven input variables comprising the geometric, soil, support system, and depth parameters as well as the output lateral wall deflections are listed in Table 7.3. Of all the wall deflection observations, 75% samples were randomly selected as the training data and the remaining 25% samples were used for testing. For the illustration of the data structures, Table 7.4 lists some sample testing data sets.

Table 7.1 Case histories of the expended database

Case no.	Soil type	Location	References
1	Stiff	Lion Yard Development, Cambridge	Ng (1992)
2		New Palace Yard Park Project, London	Burland and Hancock (1977)
4		Oxley Rise Development, Singapore	Poh et al. (1997)
5		Insurance Building, Taipei	Ou and Shiau (1998)
6		Post Office Square Parking Garage, Boston	Whittle et al. (1993)
7		Taiwan University Hospital, Chinese Taiwan	Liao and Hsieh (2002)
8		Taipei County Administration Center, Chinese Taiwan	Liao and Hsieh (2002)
9		75 State Street, Boston	Becker and Haley (1990)
10		Smith Tower, Houston	Ulrich (1989)
11		Taipei Enterprise Center, TEC	Ou et al. (1998)
12	Medium soft	Lurie Medical Building, Chicago (East Wall)	Finno and Roboski (2005)
13		Lurie Medical Building, Chicago (West Wall)	Finno and Roboski (2005)
14		Tokyo Subway Excavation Project, Japan	Miyoshi (1977)
15		HDR-4 Project for the Chicago Subway	Finno et al. (1989)
16		Oslo Subway Excavation Project	NGI (1962)
17		Embarcadero BART Zone 1, San Francisco	Clough and Buchignani (1981)
18		Metro Station: South Xizan Road, Shanghai	Wang et al. (2005)
19		Open Cut in Oslo	Peck (1969)
20		Chicago and State Street Excavation, Chicago	Finno et al. (2002)
21		Mass Rapid Transit Line, Singapore	Goh et al. (2003)
22	Soft	Excavation adjacent to Shanghai Metro Tunnels	Hu et al. (2003)
23		Excavation in downtown Chicago	Gill and Lukas (1990)

(continued)

Table 7.1 (continued)

Case no.	Soil type	Location	References
24		Peninsula Hotel Project, Bangkok	Teparaksa (1993)
25		AT&T Corporate Center, Chicago	Baker et al. (1989)
26		Chicago Museum of Science and Industry Parking	Konstantakos (2000)
27		One Market Plaza Building, San Francisco	Clough and Buchignani (1981)
28		Sheet Pile Wall Field Test, Rotterdam	Kort (2002)
29		Muni Metro Turnback Project, San Francisco	Koutsoftas et al. (2000)

Table 7.2 Summary of the excavation case histories examined (adapted from Zapata-Medina 2007)

Case no.	Soil type	H_w (m)	H_e (m)	B (m)	$EI/\gamma_w S_v^4$	$\gamma_{s,avg}$ (kN/m^3)	s_u (kPa)	δ_{hm} (mm)
1	Stiff	16.3	9.6	45.0	543	20.0	120	17.7
2		30.0	18.5	18.5	1632	20.0	170	24.1
3		33.0	20.0	63.8	1443	19.0	76.5	124.8
4		14.0	11.1	33.0	149	20.8	80	10.0
5		23.0	11.4	33.7	186	19.7	50	44.5
6		25.6	20.2	61.0	1760	20.2	91	53.6
7		27.0	15.7	140.0	2433	20.0	77.5	81.3
8		38.0	20.0	93.0	13,760	20.0	65	54.3
9		26.0	20.0	45.7	660	18.0	70	47.3
10		20.0	12.2	36.6	2748	20.1	140	14.8
11	Medium	35.0	19.7	40.0	1280	18.9	50	106.5
12		16.5	10.0	68.0	20	19.0	36	43.2
13		19.0	12.8	68.0	20	19.0	36	63.5
14		31.0	18.4	35.0	1821	19.0	47.5	62.6
15		32.0	17.0	30.0	2261	19.0	42	176.6
16		19.2	12.2	12.2	420	19.0	30	172.6
17		16.0	11.0	11.0	901	17.0	30	223.6
18		30.5	21.3	21.3	2624	18.0	44	28.3
19		38.0	20.6	22.8	510	19.0	35	48.1
20		14.0	8.5	11.0	945	19.1	27.5	228.9
21	Soft	18.3	12.2	22.0	376	19.1	20	38.1
22		31.0	16.0	20.0	3344	17.6	10	38.6
23		21.0	11.5	28.5	630	18.0	22	15.4
24		16.8	7.0	7.0	144	19.0	22.7	83.3
25		18.0	8.0	65.0	132	16.0	13.5	123.7
26		18.3	8.5	25.0	1697	19.0	21.5	37.4
27		13.7	10.3	85.0	547	19.0	45	3.6
28		30.5	11.0	11.0	1152	17.0	25	107.1
29		19.0	8.0	12.2	2	14.0	20	385.4
30		41.0	13.1	16.0	1491	16.5	27.5	48.1

(a)

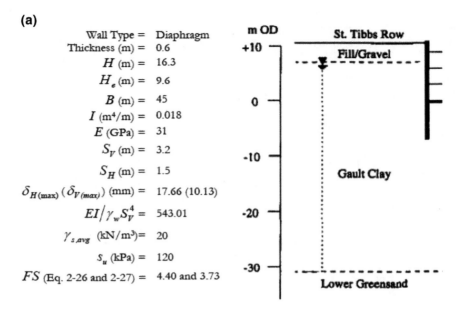

Wall Type = Diaphragm
Thickness (m) = 0.6
H (m) = 16.3
H_e (m) = 9.6
B (m) = 45
I (m⁴/m) = 0.018
E (GPa) = 31
S_V (m) = 3.2
S_H (m) = 1.5
$\delta_{H(max)}$ ($\delta_{V(max)}$) (mm) = 17.66 (10.13)
$EI/\gamma_w S_V^4$ = 543.01
$\gamma_{s,avg}$ (kN/m³)= 20
S_u (kPa) = 120
FS (Eq. 2-26 and 2-27) = 4.40 and 3.73

(b)

Depth (m)	δ_H (mm)
0.00	9.04
1.04	10.30
2.50	12.04
3.50	13.73
4.48	15.01
5.50	16.26
6.48	17.02
7.47	17.67
8.47	17.64
10.01	16.18
11.51	14.76
12.49	11.73
13.51	9.62
15.03	6.24
16.57	3.57

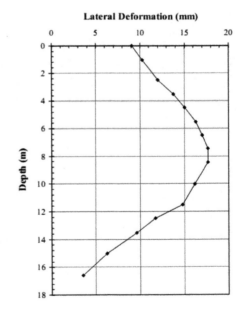

Fig. 7.1   A typical case history of case no. 1

Table 7.3 Summary of input variables and outputs

Inputs	Parameters	Physical meaning	Values or ranges
1	H_w (m)	Wall length	13.7–41.0
2	H_e (m)	Excavation depth	8.0–20.6
3	B (m)	Excavation width	11–140
4	$EI/\gamma_w S_v^4$	System stiffness	1.2–13.760
5	$\gamma_{s,avg}$ (kN/m^3)	Average unit weight of the soil	14.0–20.2
6	s_u (kPa)	Undrained shear strength	10–140
7	h (m)	Depth	0–H_w
Output δ_h (mm)		Lateral wall deflection profile	

Table 7.4 Sample testing data sets

H_w (m)	H_e (m)	B (m)	$EI/\gamma_w S_v^4$	$\gamma_{s,avg}$ (kN/m^3)	s_u (kPa)	h (m)	δ_h (mm)
16.3	9.6	45.0	543	20.0	120	7.5	17.7
16.3	9.6	45.0	543	20.0	120	12.5	11.7
33.0	20.0	63.8	1443	19.0	77	5.7	77.6
33.0	20.0	63.8	1443	19.0	77	9.6	101.0
33.0	20.0	63.8	1443	19.0	77	14.0	122.3
33.0	20.0	63.8	1443	19.0	77	18.6	120.1
33.0	20.0	63.8	1443	19.0	77	22.8	85.6
33.0	20.0	63.8	1443	19.0	77	28.6	22.2
23.0	11.4	33.7	186	19.7	50	2.0	22.9
23.0	11.4	33.7	186	19.7	50	6.7	36.1
23.0	11.4	33.7	186	19.7	50	10.0	44.1
23.0	11.4	33.7	186	19.7	50	13.4	40.2
23.0	11.4	33.7	186	19.7	50	16.6	28.2
23.0	11.4	33.7	186	19.7	50	20.0	15.6
25.6	20.2	61.0	1760	20.2	91	3.6	33.8
25.6	20.2	61.0	1760	20.2	91	8.8	49.0

7.3 The Developed MARS Model and Modeling Results

Various MARS models with a different number of BFs were developed for comparison. The optimal MARS model is determined based on the smallest relative error (defined as the ratio of the difference between the MARS predicted and target values divided by the target value, in percentage). Figures 7.2 and 7.3 show the relative error for both the location where maximum wall deflection occurs and the magnitude of the maximum wall deflection, respectively. From Figs. 7.2 and 7.3, it is obvious that the developed MARS model with 29 BFs is the most accurate.

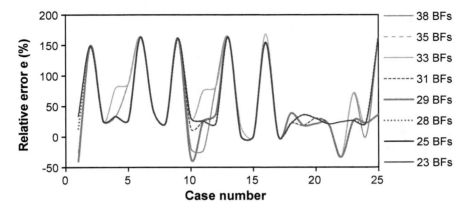

Fig. 7.2   Relative error for the location where maximum wall deflection occurs

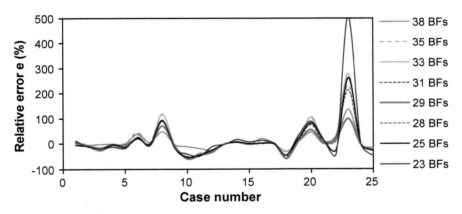

Fig. 7.3   Relative error for the magnitude of the maximum wall deflection

The training and testing predictions of δ_h at a certain depth using the optimal MARS model are shown in Fig. 7.4. It is obvious that the MARS model has been able to learn the complicated relationship between the wall deflection and the geometric, soil, support system, and depth parameters. For a majority of the data points especially those with significant wall deflections (i.e., greater than 100 mm), the relative errors fell within ±20% of the target values. Figure 7.5 plots the comparison between the target and the MARS predicted δ_{hm}, together with the corresponding location where δ_{hm} occurs. It can be observed that the developed MARS model is also capable of providing accurate estimations of the maximum wall deflection values, as well as the depth at which this maximum wall deflection occurs.

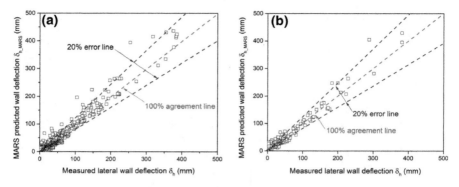

Fig. 7.4 Comparison between target and predicted δ_h

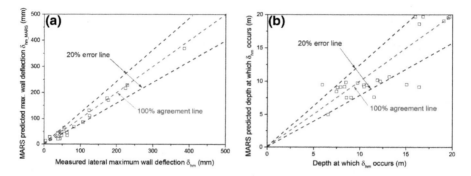

Fig. 7.5 Comparison between target and predicted: **a** δ_{hm}, and **b** depth at which δ_{hm} occurs

For a better illustration of the MARS modeling accuracy in wall deflection profile estimations, Fig. 7.6 shows the comparison of wall deflection profile between the observed and the MARS predicted for most of the cases considered. It is obvious that the developed MARS modeling results fit the observed wall deflection profile well.

Table 7.5 lists the BFs and their corresponding equation for the developed MARS model. It is observed from Table 7.5 that interactions have occurred between BFs (21 of the 29 BFs are interaction terms). The presence of interactions suggests that the built MARS model is not simply additive and that interactions play a significant role in building an accurate model for lateral wall deflection. This again indicates that MARS is capable of capturing the nonlinear and complex relationships between wall deflection profile and a multitude of influential parameters with interactions among each other without making any specific assumption about the underlying functional relationship between the input variables and the dependent response. The equation of MARS lateral wall deflection model is given by

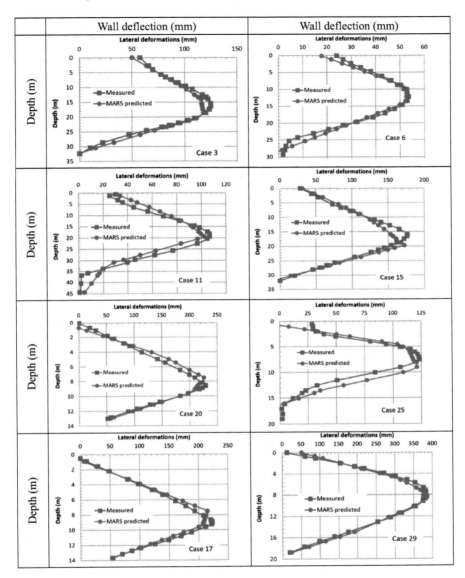

Fig. 7.6 Comparison of wall deflection profile between observed and MARS predicted

Table 7.5 Expressions of BFs for MARS model

BF	Equation	BF	Equation
1	$\max(0, B - 16)$	16	$\max(0, H_w - 31) \times \max(0, h - 19.71)$
2	$\max(0, 16 - B)$	17	$\max(0, H_w - 31) \times \max(0, 19.71 - h)$
3	$BF2 \times \max(0, 7.51 - h)$	18	$BF12 \times \max(0, h - 19.71)$
4	$BF2 \times \max(0, 420.57 - EI/\gamma_w S_v^4)$	19	$BF12 \times \max(0, 19.71 - h)$
5	$\max(0, 33 - H_w)$	20	$\max(0, H_w - 33) \times \max(0, h - 19.71)$
6	$BF5 \times \max(0, H_e - 8.5)$	21	$\max(0, H_w - 33) \times \max(0, 19.71 - h)$
7	$BF5 \times \max(0, 8.5 - H_e)$	22	$BF2 \times \max(0, h - 14.25)$
8	$\max(0, h - 9.438) \times \max(0, 8.5 - H_e)$	23	$BF2 \times \max(0, 14.25 - h)$
9	$\max(0, 9.438 - h) \times \max(0, H_e - 8.5)$	24	$\max(0, H_e - 10.3)$
10	$\max(0, 9.438 - h) \times \max(0, 8.5 - H_e)$	25	$\max(0, 10.3 - H_e)$
11	$\max(0, 31 - H_w)$	26	$BF25 \times \max(0, h - 4.53)$
12	$\max(0, H_w - 32)$	27	$BF25 \times \max(0, 4.53 - h)$
13	$\max(0, 32 - H_w)$	28	$\max(0, h - 9.438) \times \max(0, \gamma_{s,avg} - 18.9)$
14	$\max(0, h - 9.438) \times \max(0, B - 12.2)$	29	$\max(0, h - 9.438) \times \max(0, 18.9 - \gamma_{s,avg})$
15	$\max(0, h - 9.438) \times \max(0, 12.2 - B)$		

$$
\begin{aligned}
\delta_h(mm) = {} & 155.01 + 0.303 \times BF1 + 34.249 \times BF - 27.487 \times BF3 + 0.064 \\
& \times BF4 + 39.826 \times BF5 + 0.569 \times BF6 + 10.376 \times BF7 - 19.312 \\
& \times BF8 - 0.329 \times BF9 - 21.135 \times BF10 + 132.6 \times BF11 - 8.454 \\
& \times BF12 - 177.65 \times BF13 - 0.012 \times BF14 - 25.352 \\
& \times BF15 - 14.802 \times BF16 - 5.977 \times BF17 + 22.486 \\
& \times BF18 + 10.129 \times BF19 - 6.479 \times BF20 - 4.318 \\
& \times BF21 - 7.723 \times BF22 + 1.422 \times BF23 - 4.003 \\
& \times BF24 + 29.485 \times BF25 - 1.832 \times BF26 - 6.107 \\
& \times BF27 - 1.181 \times BF28 - 1.54 \times BF29
\end{aligned}
\tag{7.1}
$$

Table 7.6 ANOVA decomposition for MARS model

Function no.	GCV	STD	#Basis	Variable(s)
1	3533.72	60.85	4	H_w
2	1596.48	33.51	2	H_e
3	2987.19	53.96	2	B
4	1259.46	21.83	2	H_w, H_e
5	1248.76	21.81	6	$H_w, 7$
6	1902.33	25.51	5	H_e, h
7	1522.03	20.35	1	$B, EI/\gamma_w S_v^4$
8	1647.64	29.37	5	B, h
9	881.58	15.89	2	$\gamma_{s,avg}, h$

7.4 Parametric Sensitivity Analysis

Table 7.6 displays the ANOVA decomposition of the developed MARS model. The first column lists the ANOVA function number. The second column gives an indication of the importance of the corresponding ANOVA function, by listing the GCV score for a model with all BFs corresponding to that particular ANOVA function removed. The third column provides the standard deviation of the function. It gives an indication of its relative importance to the overall model and can be interpreted in a manner similar to the standardized regression coefficient in a linear model. The fourth column gives the number of BFs comprising the ANOVA function. The last column gives the particular input variables associated with the ANOVA function. Figure 7.7 gives the plot of the relative importance of the input variables for the MARS model, which is evaluated by the increase in the GCV value caused by removing the considered variables from the developed MARS model. The results indicate that the wall deflection at a certain depth is more sensitive to the excavation geometries and the depth compared with the soil and support system parameters. In addition, Fig. 7.7 indicates that the wall deflection profiles are not influenced by the soil undrained shear strength s_u. However, it does not mean that the maximum wall deflection value is independent of s_u.

7.5 Excavation Case History Validations

To validate the proposed MARS deflection model, two well-documented excavation case histories as listed in Table 7.7 were analyzed. It should be noted that for simplicity, only the maximum wall deflection of the two case histories is listed and compared in Table 7.7 (δ_{hm_m} denotes the measured maximum wall deflection, while δ_{hm_MARS} represents the MARS model predicted maximum wall deflection). Figure 7.8 shows the predicted lateral wall deflection at different depths/profiles versus the measured

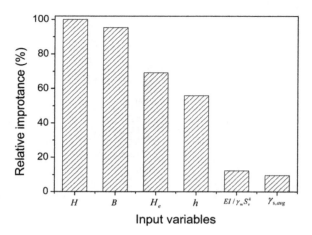

Fig. 7.7 Parameter relative importance of the developed MARS model

ones. The plot indicates that the developed MARS wall deflection profile model is able to predict reasonably well both the excavation-induced wall deflection profiles and the maximum deflection, as well as the depth at which the maximum wall deflection occurs for case histories considered.

7.6　Summary

This chapter adopted MARS algorithm to derive the predictive models for lateral wall deflections and deflection profiles induced by excavations in clays, based on an expanded database with different clay types, excavation geometries, soil, and support system parameters. Two more well-documented excavation case histories were also used to validate the proposed MARS deflection model.

Table 7.7 Summary of two excavation case histories for model validation

Case id	Case name	H_w (m)	H_e (m)	B (m)	$EI/\gamma_w S_v^4$	$\gamma_{s,avg}$ (kN/m^3)	h (m)	δ_{hm_m} (mm)	δ_{hm}_MARS (mm)	References
1	TNEC	35	19.7	41	1507	18.5	0–35	107	89	Hsieh and Ou (1998)
2	Shanghai CBD	35	20.0	65	1897	18.3	0–35	81	87	Lau et al. (2010)

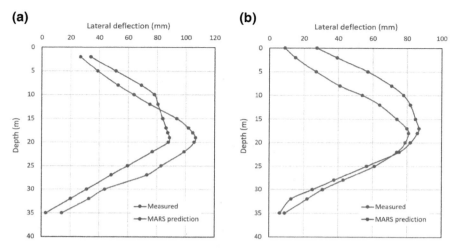

Fig. 7.8 Measured wall deflection profiles versus MARS predicted: **a** Case 1 and **b** Case 2

References

Addenbrooke TI, Potts DM, Dabee B (2000) Displacement flexibility number for multiple retaining wall design. J Geotech Geoenvironmental Eng 126(8):718–726

Baker CN Jr, Bucher SA, Baker WF Jr (1989) Complex high-rise foundation design and construction. In: Proceedings foundation engineering: current principles and practices, June 25–29, Evanston, IL, 1445–1458

Becker JM, Haley MX (1990) Up/down construction-decision making and performance. In: Proceedings of ASCE conference on design and performance of earth retaining structures, GSP 25, ASCE, New York, 170–189

Bryson L, Zapata-Medina D (2012) Method for estimating system stiffness for excavation support walls. J Geotechn Geoenvironmental Eng 138:1104–1115

Burland JB, Hancock RJR (1977) Underground car park at the house of commons: geotechnical Aspects. Struct Eng London 55(2):87–105

Clough GW, Buchignani AL (1981) Slurry walls in the San Francisco Bay Area. ASCE, Reston, pp 81–142 Preprint

Clough GW, O'Rourke TD (1990) Construction induced movements of in situ walls. Design and Performance of Earth Retaining Structures, ASCE Special Conference, Ithaca, New York, 439–470

Clough GW, Smith EM, Sweeney BP (1989) Movement control of excavation support systems by iterative design. Curr Principles Practices Foundation Eng Congress ASCE 2:869–884

Finno RJ, Roboski JF (2005) Three-dimensional responses of a tied-back excavation through clay. J Geotech Geoenvironmental Eng 131:273–282

Finno RJ, Atmatzidis DK, Perkins SB (1989) Observed performance of a deep excavation in clay. J Geotech Eng 115:1045–1064

Finno RJ, Bryson LS, Calvello M (2002) Performance of a stiff support system in soft clay. J Geotechn Geoenvironmental Eng 128:660–671

Gill SA, Lucas RG (1990) Ground movement adjacent to braced cuts. In: Proceedings of ASCE conference on design and performance of earth retaining structures, GSP 25, ASCE, New York, 471–488

Goh ATC, Wong KS, Teh CI, Wen D (2003) Pile response adjacent to braced excavation. J Geotechn Geoenvironmental Eng 129:383–386

Hashash YMA, Whittle AJ (1996) Ground movement prediction for deep excavations in soft clay. J Geotech Eng 122(6):474–486

Hsieh PG, Ou CY (1998) Shape of ground surface settlement profiles caused by excavation. Can Geotech J 35:1004–1017

Hu ZF, Yue ZQ, Zhou J, Tham LG (2003) Design and construction of a deep excavation in soft soils adjacent to the Shanghai metro tunnels. Can Geotech J 40:933–948

Konstantakos DC (2000) Measured performance of slurry walls. SM Thesis, Massachusetts Institute of Technology, Cambridge, MA.

Kort DA (2002) Steel sheet pile walls in soft soil. Ph.D. Thesis, Delft University of Technology, Delft, The Netherlands

Koutsoftas DC, Frobenius P, Wu CL, Meyersohn D, Kulesza R (2000) Deformations during cut- and cover construction of MUNI Metro Turnback project. J Geotechn Geoenvironmental Eng 126:344–359

Kung GTC, Hsiao ECL, Juang CH (2007) Evaluation of a simplified small-strain soil model for analysis of excavation-induced movements. Can Geotech J 44:726–736

Lau CS, Chiu SL, Lo KL, Chu KKN (2010) Ground response in deep excavation in soft soil in Shanghai. In: Proceedings of the 30th annual seminar geotechnical division, The Hong Kong Institution of Engineers, 149–161

Liao HJ, Hsieh PG (2002) Tied-back excavations in alluvial soil of Taipei. J Geotech Geoenvironmental Eng 128:435–441

Long M (2001) Database for retaining wall and ground movements due to deep excavations. J Geotech Geoenvironmental Eng 127:203–224

Mana AI, Clough GW (1981) Prediction of movement for braced cuts in clay. J Geotech Geoenvironmental Eng 107(6):759–777

Miyoshi M (1977) Mechanical behavior of temporary braced wall. In: Proceedings of the 6th international conference on soil mechanics and foundation engineering, Tokyo, 2(2), 655–658

Moormann C (2004) Analysis of wall and ground movements due to deep excavations in soft soil based on a new worldwide database. Soils Found 44:87–98

Ng CWW (1992) An evaluation of soil-structure interaction associated with a multi-propped excavation. Ph.D. Thesis, University of Bristol, UK

Norwegian Geotechnical Institute (1962) Measurements at a strutted excavation, Oslo Subway, Vaterland 1. Norwegian Geotechnical Institute, Technical Report 6

Ou CY, Chiou DC (1998) Analysis of the corner effect on excavation behaviors. Can Geotech J 35(3):532–540

Ou CY, Liao JT, Lin HD (1998) Performance of diaphragm wall constructed using top-down method. J Geotech Geoenvironmental Eng 124:798–808

Peck RB (1969) Deep excavations and tunneling in soft ground. In: 7th ICSMFE, state of the art report, Mexico, 225–290

Poh TY, Wong IH, Chandrasekaran B (1997) Performance of two propped diaphragm walls in stiff residual soils. J Perform Constructed Facil 11:190–199

Teparaksa W (1993) Behavior of deep excavations using sheet pile bracing system in soft Bangkok clay. In: Third international conference on case histories in geotechnical engineering, June 1–6. St. Louis, USA, 745–750

Ulrich EJ Jr (1989) Tieback supported cuts in over-consolidated soils. J Geotech Eng 115:521–545

Wang ZW, Ng CWW, Liu GB (2005) Characteristics of wall deflections and ground surface settlements in Shanghai. Can Geotech J 42:1243–1254

Whittle AJ, Hashash YMA, Whitman RV (1993) Analysis of deep excavation in Boston. J Geotech Eng 119:69–90

Wong KS, Broms BB (1989) Lateral wall deflections of braced excavation in clay. J Geotech Eng 115(6):853–870

Zapata-Medina DG (2007) Semi-empirical method for designing excavation support systems based on deformation control. MSc thesis, University of Kentucky, Lexington, USA

Zhang WG, Goh ATC, Xuan F (2015a) A simple prediction model for wall deflection caused by braced excavation in clays. Comput Geotech 63:67–72

Zhang WG, Goh ATC, Zhang YM, Chen YM, Xiao Y (2015b) Assessment of soil liquefaction based on capacity energy concept and multivariate adaptive regression splines. Eng Geol 188:29–37

Chapter 8
MARS Use for Determination of EPB Tunnel-Related Maximum Surface Settlement

A major consideration in urban tunnel design is to estimate the ground movements and surface settlements associated with the tunneling operations. Excessive ground movements may result in damage to adjacent buildings and utilities. Numerous empirical and analytical solutions have been proposed to relate the shield tunnel characteristics and surface/subsurface deformation. Also, numerical analyses, either 2D or 3D, have been used for such tunneling problems. However, substantially fewer approaches have been developed for earth pressure balance (EPB) tunneling. Based on instrumented data on the ground deformation and shield operation from three separate EPB tunneling projects in Singapore, this paper utilizes MARS approach to establish relationships between the maximum surface settlement and the major influencing factors.

8.1 Background

Due to the population growth and urbanization, there is an increasing demand for the construction of tunnels for mass rapid transportation services. Mechanized excavations using earth pressure balance (EPB) have been applied successfully around the world, especially in urban environments where there is less surface space available. Ground movements and surface settlements associated with shield tunneling operations are a major concern in the design of tunnels in urban areas, as excessive movements can damage nearby buildings and utilities.

Generally, there are three major classes of the ground settlement prediction models: empirical and analytical methods, numerical models, and the soft computing approaches driven by the measured data from case histories.

In the early stage of the design, the designer bases his estimations on previous experience and may use simple empirical equations, such as those described by Peck (1969), Attewell and Farmer (1974), Atkinson and Potts (1977), Yoshikoshi et al. (1978), O'Reilly and New (1982), Attewell et al. (1986), Mair et al. (1993), or simple equations based on the theory of elasticity (Uriel and Sagaseta 1989). Recently,

© Science Press and Springer Nature Singapore Pte Ltd. 2019
W. Zhang, *MARS Applications in Geotechnical Engineering Systems*,
https://doi.org/10.1007/978-981-13-7422-7_8

several works have extended Peck's formula to model the surface settlement of twin- and quadruple-tube tunnels (Chen et al. 2012; Gui and Chen 2013). Analytical solutions were developed by Segaseta (1987), Verruijt and Booker (1996), Loganathan and Poulos (1998), Bobet (2001), Chou and Bobet (2002), Park (2005). However, the final design generally requires more accurate stress and deformation analyses, and finite difference or finite element methods may therefore usually be used, as discussed by Rowe and Lee (1992). Moreover, there are doubts as to the accuracy of some of these methods since they fail to take into account all the relevant factors, which include many shield operational parameters, that all concurrently influence ground settlement (Kim et al. 2001).

With the rapid development of computer technology and numerical algorithms, numerical models have been used widely in tunneling projects (Rowe and Lee 1992; Swoboda and Abu-Krisha 1999; Addenbrooke and Potts 2001; Mroueh and Shahrour 2002; Kasper and Meschke 2006a, b; Callari 2004; Ng and Lee 2005; Ocak 2009; Chakeri et al. 2010; Ercelebi et al. 2011; Lambrughi et al. 2012). Recently, Fahimi-far and Zareifard (2013) proposed an analytical–numerical approach to simulate the response of tunnels under different hydromechanical conditions. Chen et al. (2014) presented a full 3D seismic analysis on a water conveyance tunnel. Talebinejad et al. (2014) investigated the surface and subsurface displacements due to multiple tunnel excavations in the Tehran region using a full 3D finite difference analysis, with special attention paid to the effect of subsequent tunneling on the support system. Nevertheless, the implementation of a numerical model is relatively complex, particularly when mechanized processes of shield excavation are considered. In addition, the predictive performance of numerical models depends significantly on the model describing the soil behavior (Karakus and Fowell 2003). Detailed information on soil properties, which is required for simulation, is lacking or unavailable in many cases, and thus building a practical constitutive soil model for tunneling-induced settlement prediction is rather difficult (Karakus and Fowell 2005). Moreover, the formulation of conventional finite elements has difficulties in capturing the onset of strain localization and its propagation from the tunnel up to the ground surface, and, considering the coupled relationship between the soil response and fluid flow, is of great importance (Callari 2004; Callari et al. 2010). Even with powerful numerical tools, a significant effort and computational time are needed to correlate ground movements with these various parameters.

During the past decade, artificial neural networks (ANNs) have also been used as an alternative method for solving this complex nonlinear problem. Most of the ANN-based analyses were implemented by extracting the relationship between the influencing parameters, such as the shielding operation factors, tunnel depth and diameter, soil properties, and the induced settlement. ANN has been adopted successfully for predicting the maximum surface settlement in a number of tunneling projects (Shi et al. 1998; Kim et al. 2001; Suwansawat and Einstein 2006; Santos and Celestino 2008; Goh and Hefney 2010; Xu and Xu 2011; Pourtaghi and Lotfollahi-Yaghin 2012; Ocak and Seker 2013; Mohammadi et al. 2015). In all these projects, the neural networks were trained and tested using instrumented data as well as data from the tunnel operational parameters and geological parameters. However, when

Table 8.1 Summary of the computing methods and their use in ground settlement estimation

Computing method	References
Relevance vector machines (RVMs)	Wang et al. (2013)
Gaussian processes (GP)	Ocak and Seker (2013)
Partial least squares regression method	Bouayad et al. (2015)
Decision tree method	Dindarloo and Siamilrdemoosa (2015)
Adaptive neuro-fuzzy inference system (ANFIS)	Hou et al. (2009)
Gene expression programming (GEP)	Ahangari et al. (2015)

using ANNs, it is difficult to determine the optimal network architecture, and ANNs often suffer the problem of poor generalization and model interpretation.

SVMs, which are based on statistical learning, have also been applied successfully to predicting ground settlement due to underground excavations (Neaupane and Adhikari 2006; Samui 2008; Yao et al. 2010; Jiang et al. 2011; Wang et al. 2012; Ocak and Seker 2013). SVMs also have drawbacks such as the determination of model parameters (e.g., the penalty weight C and the insensitivity parameter e), relative high model complexity, and kernel function restrictions.

Other computing methods and their ground settlement estimation applications are listed in Table 8.1.

This chapter adopts the MARS method to map all influencing parameters including the operation parameters, the cover depth, and the ground conditions to surface settlements, based on three separate mass rapid transit projects in Singapore.

8.2 Ground Conditions

Major works for the construction of mass rapid transit tunnels in Singapore have been described in a number of publications (Hulme and Burchell 1999; Izumi et al. 2000; Shirlaw et al. 2003). The four major soil types encountered in this study were:

1. Kallang formation. This is comprised of near normally consolidated marine clay as well as loose fluvial sands and moderately stiff fluvial clay.
2. Old Alluvium. This is comprised mainly of dense alluvial silty sands and clays.
3. Fort Canning Boulder Bed. This is comprised of a colluvial deposit of strong to very strong quartzite boulders in a hard clayey silt matrix.
4. Jurong formation. This is comprised primarily of residual soils of clayey silt and sandy clay of medium plasticity and clayey to silty sand.

The tunnel settlement data used in this paper were obtained from three separate mass rapid transit projects of the Circle Line in Singapore. The locations of the projects are shown in Fig. 8.1. Figure 8.2 shows the soil profiles of the three projects. Twin-bored tunnel drives using EPB machines were used (Table 8.2). Precast concrete

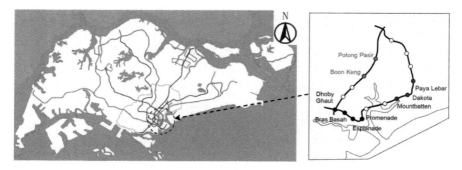

Fig. 8.1 Map showing the project locations (Google map http://www.maps.google.com)

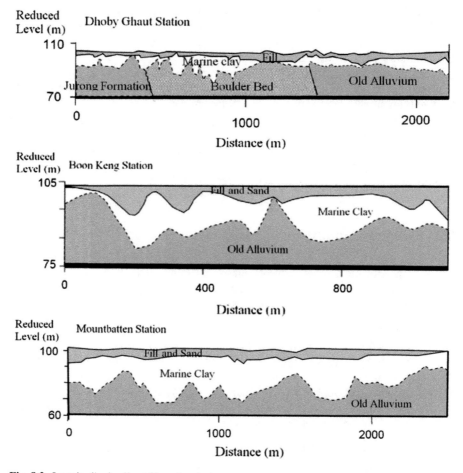

Fig. 8.2 Longitudinal soil profiles of projects

Table 8.2 Summary of tunnel project details

Contract	C705	C823	C825
TBM manufacture	Hitachi-Zosen	Hitachi-Zosen	Herrenknecht
Drive length (km)	1.3	3.3	1.5
Tunnel drive (no.)	2	2	2
Outside diameter (m)	6.44	6.63	6.58
Internal diameter (m)	5.8	5.8	5.8
Stations	Boon Keng and Potong Pasir	Mountbatten, Dakota and Paya Lebar	Dhoby Ghaut, Bras Basah, Esplanade and Promenade
Geology (General description)	Mostly Old Alluvium with marine clay	Fill overlying Kallang formation and Old Alluvium	Soft marine clay, Old Alluvium, Fort Canning Boulder bed, Jurong formation

Table 8.3 Statistical summary of settlement data sets

Parameter and description	Min	Max	Ave	Stdev
Cover H (m)	8.5	30	17.5	4.3
Advance rate AR (mm/min)	9.5	52.1	30.8	10.9
Earth pressure EP (kPa)	11	370	193.6	81.5
Mean SPT above crown level $S1$ (blows/300 mm)	0.66	80.33	27.9	28.2
Mean tunnel SPT $S2$ (blows/300 mm)	0	100	57.0	41.8
Mean moisture content MC (%)	5.95	66.48	27.1	18.7
Mean soil elastic modulus E (MPa)	5	120	72.9	50.8
Grout pressure GP (kPa)	27.7	700	258.6	154.9
Surface settlement (mm)	0.2	98.5	13.6	17.0

tunnel linings of 5.8 m internal diameter were used throughout. Surface settlement points were installed at approximately 25 m intervals along the tunnel alignment. A total of 148 settlement patterns were obtained from the three projects.

8.3   Data Sets and the Influential Parameters

A total of 148 instrumented sections of settlement data (patterns) were obtained from the three projects. The denoted symbols, descriptions, and ranges of the various parameters are shown in Table 8.3.

There are actually numerous factors that influence the surface settlement; these can be subdivided into three major categories (Suwansawat and Einstein 2006): (1) tunnel geometry, (2) geological conditions, and (3) EPB operation factors.

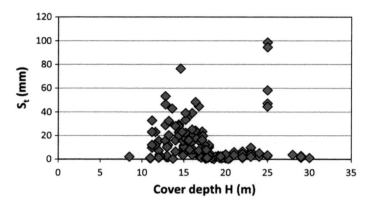

Fig. 8.3 Relationship between H and surface settlement S_t

8.3.1   Tunnel Geometry

The two important geometrical factors that influence surface settlement are tunnel diameter and tunnel depth, H (measured from the tunnel crown to the ground surface). However, since the internal diameter of the tunnels for all three projects was 5.8 m, this parameter was omitted from the MARS and neural network analysis. Figure 8.3 shows the tunnel cover depth H versus the measured maximum surface settlement S_t.

8.3.2   Geological Conditions

The geological factors used as inputs for the analyses were $S1$, the mean SPT N value (standard penetration test) of the soil layers above crown level up to ground surface; $S2$, the average of the SPT N values at the crown, middle, and invert levels; MC, the average moisture content of the soil layer driven through by the tunnel machine; and E, the average modulus of elasticity of the soil layer driven through by the tunnel machine. As the depth of the groundwater table did not vary significantly in these three projects, it was omitted as one of the input parameters. Figures 8.4, 8.5, 8.6, and 8.7 plot the relationship between the measured maximum surface settlement S_t and $S1$, $S2$, MC, and E, respectively.

8.3.3   EPB Operation Factors

The EPB operational factors used as inputs were AR, the tunnel advance rate; EP, the EPB earth (face) pressure; and GP, the grout pressure used for injecting grout

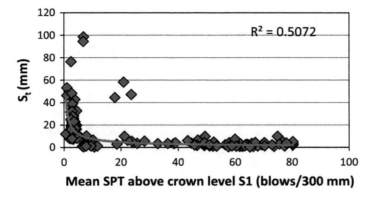

Fig. 8.4   Relationship between $S1$ and surface settlement S_t

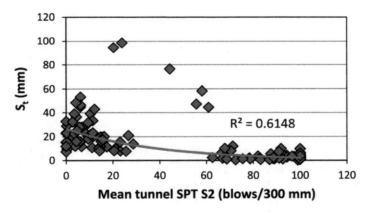

Fig. 8.5   Relationship between $S2$ and surface settlement S_t

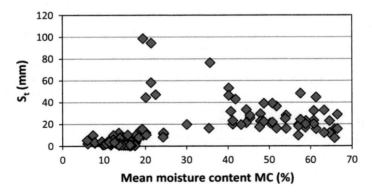

Fig. 8.6   Relationship between MC and surface settlement S_t

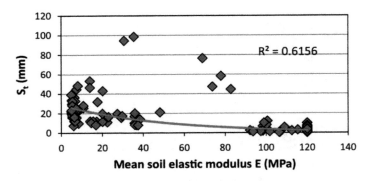

Fig. 8.7 Relationship between E and surface settlement S_t

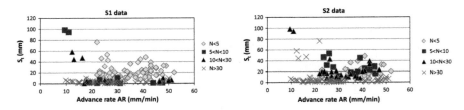

Fig. 8.8 Relationship between $S1$, $S2$, tunnel advance rate (AR) and surface settlement

into the tail void. Figure 8.8a shows the tunnel advance rate AR versus the measured maximum surface settlement as a function of $S1$. Figure 8.8b shows the plot of the tunnel advance rate AR versus the measured maximum surface settlement as a function of $S2$. In general, while it can be seen that the surface settlements are less than 20 mm for $S1$ and $S2$ with N values greater than 30, it is difficult to establish any well-defined relationship between the surface settlement, the advance rate, $S1$ and $S2$. Figure 8.9a, b shows the plot of EP versus S_t as a function of $S1$ and $S2$, respectively. Figure 8.10a, b shows the plot of GP versus S_t as a function of $S1$ and $S2$, respectively. Generally, these figures show a considerable scatter in the data. Hence, a multivariate adaptive regression spline algorithm, as a statistical, nonlinear, and nonparametric regression method for fitting the relationship between a set of input variables and dependent variables, was used to determine the relationship between the surface settlement, and the geological and EPB operational factors. As will be shown in the following section, the MARS approach is capable of learning from the training examples and can capture the nonlinear and complex interactions among the various inputs and the surface settlement.

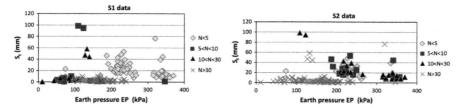

Fig. 8.9 Relationship between $S1$, $S2$, earth pressure (EP) and surface settlement

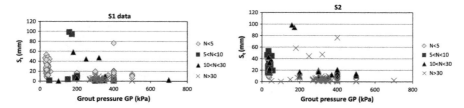

Fig. 8.10 Relationship between $S1$, $S2$, grout pressure (GP) and surface settlement

8.4 The Developed MARS Model and Modeling Results

The MARS model used to model the tunnel settlement consisted of eight inputs that represented the tunnel geometry, geological conditions, and the EPB operation factors as listed in Table 8.3. From the 148 data sets, a total of 115 sets of settlement data samples (patterns) were selected randomly as the training data, and the remaining 33 data samples were used for testing the validity of the developed MARS model. A data sample refers to a set of input data and associated measured settlement corresponding to an instrumented section. Since, for most practical cases, the serviceability limit is less than 30 mm, the testing samples that were less than 30 mm were randomly selected, while ensuring that the training and testing samples are statistically consistent. The full database is referred to Goh et al. (2018).

For the MARS model, the logarithmic values of parameters EP, E, GP, and S_t were used, as it was found that this substantially improves the MARS's training process. The tunnel settlement analysis using MARS adopted 16 BFs of linear spline functions with the second-order interaction. The plots of the MARS predicted versus measured settlement values are shown in Fig. 8.11. For comparison, the same sets of training and testing patterns were analyzed using the ANN method. Based on trial and error, the optimal ANN model consists of five hidden neurons. The results indicated a fairly high coefficient of determination ($R2$) between the actual and predicted settlement values of 0.906 and 0.721 for the training and testing samples, respectively, compared with the fitting results of 0.873 and 0.689 using ANN at the same time. The mean average error (MAE) was 3.24 mm for the training samples and 3.58 mm for the testing samples. These values are smaller than those of 4.19 mm for the training patterns and 3.64 mm for the testing patterns obtained through ANN.

Fig. 8.11 Comparison of measured and MARS-predicted S_t

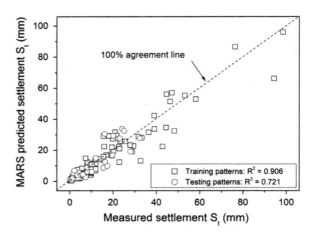

Table 8.4 lists the BFs and their corresponding equations. It is noted from Table 8.4 that of the 16 BFs, 13 (i.e., all except BF1, BF5, and BF10) with interaction terms are integrated in this model, indicating that the model is not simply additive and that interaction terms play a significantly important role. The mathematical expression for this interpretable MARS model is given by

$$
\begin{aligned}
\log(S\_t) = {} & 1.4764 - 4.033 \times \mathrm{BF1} - 13.466 \times \mathrm{BF2} + 3.885 \times \mathrm{BF3} + 0.756 \times \mathrm{BF4} \\
& - 0.062 \times \mathrm{BF5} - 0.357 \times \mathrm{BF6} - 12.328 \times \mathrm{BF7} - 2.941 \times \mathrm{BF8} + 0.082 \times \mathrm{BF9} \\
& + 0.422 \times \mathrm{BF10} - 1.768 \times \mathrm{BF11} - 3.813 \times \mathrm{BF14} - 88.602 \times \mathrm{BF13} + 0.468 \times \mathrm{BF14} \\
& - 0.0036 \times \mathrm{BF15} + 0.026 \times \mathrm{BF16}
\end{aligned} \tag{8.1}
$$

8.5   Parametric Sensitivity Analyses

Table 8.5 displays the ANOVA decomposition of the developed MARS model. The first column in Table 8.5 lists the ANOVA function number. The second column gives an indication of the importance of the corresponding ANOVA function, by listing the GCV score for a model with all BFs corresponding to that particular ANOVA function removed. This GCV score can be used to evaluate whether the ANOVA function is making an important contribution to the model, or whether it just marginally improves the global GCV score. The third column provides the standard deviation of this function. The fourth column gives the number of BFs comprising the ANOVA function. The last column gives the particular input variables associated with the ANOVA function. Figure 8.12 gives the plots of the relative importance of the input variables for the developed MARS model, which is evaluated by the increase in the GCV value caused by removing the variables considered from the developed MARS model. It can be observed that EP is the most important parameter, followed by mean MC, AR, and GP.

Table 8.4 Basis functions and corresponding equations of MARS model for settlement prediction

Basis function	Equation
BF1	$\max(0, \log(EP) - 2.0492)$
BF2	$\max(0, \log(E) - 1.8921) \times \max(0, \log(GP) - 2.1761)$
BF3	$BF1 \times \max(0, AR - 21.2)$
BF4	$BF1 \times \max(0, 21.2 - AR)$
BF5	$\max(0, MC - 35.7)$
BF6	$\max(0, \log(E) - 1.8921) \times \max(0, H - 23)$
BF7	$\max(0, 2.0492 - \log(EP)) \times \max(0, \log(GP) - 2.5065)$
BF8	$\max(0, 2.0492 - \log(EP)) \times \max(0, 2.5065 - \log(GP))$
BF9	$\max(0, 35.7 - MC) \times \max(0, 2.0622 - \log(E))$
BF10	$\max(0, MC - 17.5)$
BF11	$\max(0, 1.8921 - \log(E)) \times \max(0, \log(GP) - 1.6628)$
BF12	$BF10 \times \max(0, \log(EP) - 1.9469)$
BF13	$\max(0, 1.8921 - \log(E)) \times \max(0, 2.0362 - \log(EP))$
BF14	$\max(0, 22 - H) \times \max(0, \log(GP) - 2.4771)$
BF15	$\max(0, 22 - H) \times \max(0, 25.67 - S1)$
BF16	$\max(0, 22 - H) \times \max(0, 3.25 - S1)$

Table 8.5 ANOVA decomposition of MARS model

Function no.	GCV	STD	#basis	Variable(s)
1	42.591	3.966	1	EP
2	56.969	4.653	2	MC
3	36.700	2.790	2	H and $S1$
4	7.600	1.107	1	H and E
5	20.900	2.227	1	H and GP
6	92.566	5.905	2	AR and EP
7	31.720	2.562	1	EP and MC
8	4.610	0.919	1	EP and E
9	47.800	4.231	2	EP and GP
10	12.800	1.752	1	MC and E
11	11.260	1.643	2	E and GP

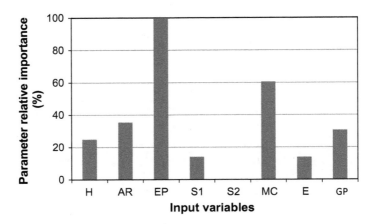

Fig. 8.12 Relative importance of the input variables selected in the MARS model

To further validate the MARS EPB tunnel-related settlement model, a parametric analysis was performed, with the aim of studying the effect of TBP shield operation factors on the induced maximum ground settlement. This parametric sensitivity analysis investigates the response of $Log(S_t)$ predicted by the MARS model to a set of hypothetical input data generated over the ranges of the minimum and maximum data sets. One input variable was changed each time within its range, while the others were kept at the average values of their entire data sets. As suggested by Alavi et al. (2011), a set of synthetic data for the single varying parameter was generated by increasing the value of this in increments. These values were presented to the MARS prediction model and $Log(S_t)$ was calculated. This procedure was repeated using another variable until the responses of the models were tested for all of the predictor variables (Alavi et al. 2011). Figure 8.13a–c presents the effects of the $Log(S_t)$ predictions to the variations of AR, EP, and GP, respectively. Figure 8.13a confirms the findings of Santos and Celestino (2008) that the maximum surface settlement value increases with increase of AR and reaches a plateau. Subsequently, it then decreases as the AR rises continuously. It is obvious from Fig. 8.13b, c that the maximum surface settlement decreases as the EP and GP become larger.

8.6 Summary

The developed MARS model can be used for determination of EPB tunnel-related settlement in similar ground. The interaction terms integrated in this model (Eq. 8.1; Table 8.4) indicate that the interactions between the geological conditions and EPB operation factors play a significant role in determining the maximum settlement. Also, geological conditions essentially influence the set-up of EPB operational fac-

Fig. 8.13 Parametric
analysis of the EPB tunnel
related settlement MARS
model

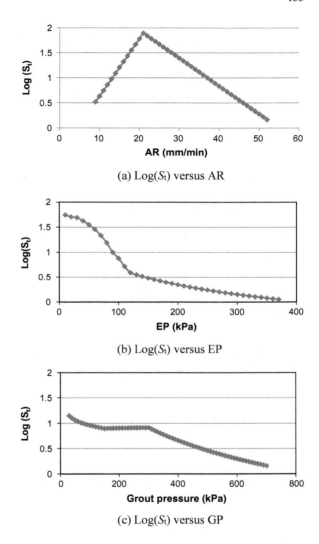

(a) Log(S_t) versus AR

(b) Log(S_t) versus EP

(c) Log(S_t) versus GP

tors. However, it should be noted that there are still some limitations with regard to
use of the built MARS model:

(1) Since the built MARS model makes predictions based on the knot values and
the basis functions, interpolations between the knots of design input variables
are more accurate and reliable than extrapolations. Consequently, it is not recommended that the model be applied for values of input parameters beyond the
specific ranges in this study.

(2) The diameter of the shield tunnel and the other operational factors such as segment thickness and length, cutter head rotational speed, and maximum torque should also be taken into account.

(3) It should also be pointed out that the groundwater level change is excluded in this analysis since the depth of groundwater table during construction did not vary significantly in these three projects. In general, minimal lowering of the groundwater is permitted for the construction of mass rapid tunnels in Singapore.

References

Addenbrooke TI, Potts DM (2001) Finite element analysis of St. James Park greenfield reference site. In: Burland JB, Standing JR, Jardine FM (eds) Building response to tunnelling, vol 1. Thomas Telford, London, pp 177–194

Ahangari K, Moeinossadat SR, Behnia D (2015) Estimation of tunnelling-induced settlement by modern intelligent methods. Soils Found 55(4):737–748

Alavi AH, Ameri M, Gandomi AH, Mirzahosseini MR (2011) Formulation of flow number of asphalt mixes using a hybrid computational method. Constr Build Mater 25:1338–1355

Atkinson JH, Potts DM (1977) Subsidence above shallow tunnels in soft ground. In: Proceedings of ASCE Geotechnical Engineering Division, 59–64

Attewell PB, Farmer IW (1974) Ground deformations resulting from shield tunnelling in London Clay. Can Geotech J 11:380–395

Attewell PB, Yeates J, Selby AR (1986) Soil Movements Induced by Tunneling. Chapman & Hall, New York

Bobet A (2001) Analytical solutions for shallow tunnels in saturated ground. J Eng Mech 12:1258–1266

Bouayad D, Emeriault F, Maza M (2015) Assessment of ground surface displacements induced by an earth pressure balance shield tunneling using partial least squares regression. Environ Earth Sci 73:7603–7616

Callari C (2004) Coupled numerical analysis of strain localization induced by shallow tunnels in saturated soils. Comput Geotech 31:193–207

Callari C, Armero F, Abati A (2010) Strong discontinuities in partially saturated poroplastic solids. Comput Methods Appl Mech Eng 199:1513–1535

Chakeri H, Hasanpour R, Hindistan MA, Unver B (2010) Analysis of interaction between tunnels in soft ground by 3D numerical modeling bulletin of engineering geology and the environment. Springer, Berlin

Chen SL, Gui MW, Yang MC (2012) Applicability of the principle of superposition in estimating ground surface settlement of twin- and quadruple-tube tunnels. Tunn Undergr Space Technol 28:135–149

Chen Z, Yu H, Yuan Y (2014) Full 3D seismic analysis of a long distance water conveyance tunnel. Struct Infrastruct Eng 10:128–140

Chou WL, Bobet A (2002) Predictions of ground deformations in shallow tunnels in clay. Tunn Undergr Space Technol 17:3–19

Dindarloo SR, Siami-Irdemoosa E (2015) Maximum surface settlement based classification of shallow tunnels in soft ground. Tunn Undergr Space Technol 49:320–327

Ercelebi SG, Copur H, Ocak I (2011) Surface settlement predictions for Istanbul Metro tunnels excavated by EPB-TBM. Earth-Sci Rev 62:357–365

Fahimifar A, Zareifard MR (2013) A new elasto-plastic solution for analysis of underwater tunnels considering strain-dependent permeability. Struct Infrastruct Eng 10:1432–1450

Goh ATC, Hefney AM (2010) Reliability assessment of EPB tunnel-related settlement. Int J Geomech Eng 2(1):57–69

Goh ATC, Zhang WG, Zhang YM, Xiao Y, Xiang YZ (2018) Determination of EPB tunnel-related maximum surface settlement: A Multivariate adaptive regression splines approach. Bull Eng Geol Env 77:489–500

Gui MW, Chen SL (2013) Estimation of transverse ground surface settlement induced by DOT shield tunneling. Tunn Undergr Space Technol 33:119–130

Hou J, Zhang MX, Tu M (2009) Prediction of surface settlements induced by shield tunneling: an ANFIS model. In: Huang Liu (ed) Geotechnical Aspects of Underground Construction in Soft Ground—Ng. Taylor & Francis Group, London, pp 551–554

Hulme TW, Burchell AJ (1999) Tunneling projects in Singapore: an overview. Tunn Undergr Space Technol 14(4):409–418

Izumi C, Khatri NN, Norrish A, Davies R (2000) Stability and settlement due to bored tunneling for LTA, NEL. In: Proceedings of international conference on tunnels and underground structures, Singapore, 555–560

Jiang AN, Wang SY, Tang SL (2011) Feedback analysis of tunnel construction using a hybrid arithmetic based on support vector machine and particle swarm optimisation. Autom Constr 20:482–489

Karakus M, Fowell RJ (2003) Effects of different tunnel face advance excavation on the settlement by FEM. Tunn Undergr Space Technol 18:513–523

Karakus M, Fowell RJ (2005) Back analysis for tunnelling induced ground movements and stress redistribution. Tunn Undergr Space Technol 20:514–524

Kasper T, Meschke G (2006a) A numerical study of the effect of soils and grout materiel properties and cover depth in shield tunneling. Comput Geotech 33(4–5):234–247

Kasper T, Meschke G (2006b) On the influence of face pressure, grouting pressure and TBM design in soft ground tunneling. Tunn Undergr Space Technol 21:160–171

Kim CY, Bae GJ, Hong SW, Park CH, Moon HK, Shin HS (2001) Neural network based prediction of ground settlements due to tunneling. Comput Geotech 28(6–7):517–547

Lambrughi A, Medina Rodriguez L, Castellanza R (2012) Development and validation of a 3D numerical model for TBM-EPB mechanised excavations. Comput Geotech 40:97–113

Loganathan N, Poulos HG (1998) Analytical prediction for tunneling-induced ground movements in clay. J Geotechn Geoenvironmental Eng 124(9):846–856

Mair RJ, Taylor RN, Bracegirdle A (1993) Subsurface settlement profiles above tunnels in clays. Geotechnique 43(2):315–320

Mohammadi SD, Naseri F, Alipoor S (2015) Development of artificial neural networks and multiple regression models for the NATM tunnelling-induced settlement in Niayesh subway tunnel, Tehran. Bull Eng Geol Environ 74:827–843

Mroueh H, Shahrour I (2002) Three-dimensional finite element analysis of the interaction between tunneling and pile foundations. Int J Numer Anal Methods Geomechanics 26:217–230

Neaupane KM, Adhikari NR (2006) Prediction of tunnelinginduced ground movement with the multi-layer perceptron. Tunn Undergr Space Technol 21:151–159

Ng CW, Lee GTK (2005) Three-dimensional ground settlements and stress-transfer mechanisms due to open-face tunneling. Can Geotech J 42:1015–1029

Ocak I (2009) Environmental effects of tunnel excavation in soft and shallow ground with EPBM: the case of Istanbul. Environ Earth Sci 59(2):347–352

Ocak I, Seker SE (2013) Calculation of surface settlements caused by EPBM tunneling using artificial neural network, SVM, and Gaussian processes. Environ Earth Sci 70:1263–1276

O'Reilly MP, New BM (1982) Settlement above tunnels in the United Kingdom, their magnitude and prediction. In Proceedings of Tunnelling 82, Brighton. London, 173–181

Park KH (2005) Analytical solution for tunneling-induced ground movement in clays. Tunn Undergr Space Technol 20(3):249–261

Peck RB (1969) Deep excavations and tunneling in soft ground. In: 7th ICSMFE, state of the art report, Mexico, 225–290

Pourtaghi A, Lotfollahi-Yaghin MA (2012) Wavenet ability assessment in comparison to ANN for predicting the maximum surface settlement caused by tunneling. Tunn Undergr Space Technol 28:257–271

Rowe RK, Lee KM (1992) An evaluation of simplified techniques for estimating three dimensional undrained ground movements due to tunneling in soft soils. Can Geotech J 29:39–52

Samui P (2008) Support vector machine applied to settlement of shallow foundations on cohesion-less soils. Comput Geotech 35:419–427

Santos OJ Jr, Celestino TB (2008) Artificial neural networks of Sao Paulo subway tunnel settlement data. Tunn Undergr Space Technol 23:481–491

Segaseta C (1987) Analysis of undrained soil deformation due to ground loss. Geotechnique 37:301–320

Shi J, Ortigao JAR, Bai J (1998) Modular neural networks for predicting settlements during tun-neling. J Geotechn Geoenvironmental Eng ASCE 124(5):389–395

Shirlaw JN, Ong JCW, Rosser HB, Tan CG, Osborne NH, Heslop PE (2003) Local settlements and sinkholes due to EPB tunneling. Geotechn Eng ICE 156(GE4):193–211

Suwansawat S, Einstein HH (2006) Artificial neural networks for predicting the maximum surface settlement caused by EPB shield tunneling. Tunn Undergr Space Technol 21(2):133–150

Swoboda G, Abu-Krisha A (1999) Three dimensional numerical modeling for TBM tunneling in consolidated clay. Tunn Undergr Space Technol 14(3):327–333

Talebinejad A, Chakeri H, Moosavi M, Özçelik Y, Ünver B, Hindistan M (2014) Investigation of surface and subsurface displacements due to multiple tunnels excavation in urban area. Arab J Geosci 7:3913–3923

Uriel AO, Sagaseta C (1989) Selection of design parameters for underground construction. In: Proceedings of 12th ICSMFE, Rio de Janeiro, Brazil, Balkema 4, 2521–2551

Verruijt A, Booker JR (1996) Surface settlement due to deformation of a tunnel in an elastic half place. Geotechnique 46(4):753–756

Wang F, Gou BC, Qin YW (2013) Modeling tunneling-induced ground surface settlement devel-opment using a wavelet smooth relevance vector machine. Comput Geotech 54:125–132

Wang DD, Qiu GQ, Xie WB, Wang Y (2012) Deformation prediction model of surrounding rock based on GA-LSSVM-markov. Nat Sci 4(2):85–90

Xu J, Xu Y (2011) Grey correlation-hierarchical analysis for metro-caused settlement. Environ Earth Sci 64(5):1246–1256

Yao BZ, Yang CY, Yao JB, Sun J (2010) Tunnel surrounding rock displacement prediction using support vector machine. Int J Comput Int Sys 3(6):843–852

Yoshikoshi W, Osamu W, Takagaki N (1978) Prediction of ground settlements associated with shield tunnelling. Soils Found 18:47–59

Chapter 9
MARS Use in HP-Pile Drivability Assessment

Piles are long, slender structural elements used to transfer the loads from the super-structure through weak strata onto stiffer soils or rocks. For driven piles, the impact of the piling hammer induces compression and tension stresses in the piles, as shown in Fig. 9.1. Hence, an important design consideration is to check that the strength of the pile is sufficient to resist the stresses caused by the impact of the pile hammer.

Due to its complexity, pile drivability lacks a precise analytical solution with regard to the phenomena involved. In situations where measured data or numerical hypothetical results are available, neural networks stand out in mapping the non-

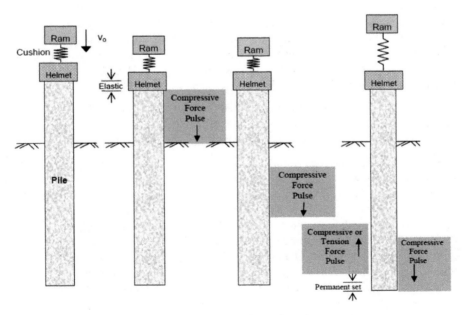

Fig. 9.1 Schematic plot of piling-induced compression and tension stresses

© Science Press and Springer Nature Singapore Pte Ltd. 2019
W. Zhang, *MARS Applications in Geotechnical Engineering Systems*,
https://doi.org/10.1007/978-981-13-7422-7_9

linear interactions and relationships between the system's predictors and dependent responses. In addition, unlike most computational tools, no mathematical relationship assumption between the dependent and independent variables has to be made. Nevertheless, neural networks have been criticized for their long trial-and-error training process since the optimal configuration is not known a priori. This chapter investigates the use of MARS, as an alternative to BPNN, to approximate the relationship between the inputs and dependent response, and to mathematically interpret the relationship between the various parameters. In this chapter, the BPNN and MARS models are developed for assessing pile drivability in relation to the prediction of the maximum compressive stresses (MCSs), maximum tensile stresses (MTSs), and blow per foot (BPF). A database of more than four thousand piles is utilized for model development and comparative performance between BPNN and MARS predictions.

9.1 Background

Piles are long, slender structural elements used to transfer the loads from the superstructure through weak strata onto stiffer soils or rocks. The selection of the type of pile depends on the type of structure, the ground conditions, the durability (e.g., to corrosion), and the installation costs. For driven piles, the impact of the piling hammer induces compression and tension stresses in the piles. Hence, an important design consideration is to check that the strength of the pile is sufficient to resist the stresses caused by the impact of the pile hammer. One common method of calculating driving stresses is based on the stress-wave theory (Smith 1960) which involves the discrete idealization of the hammer–pile–soil system. As the conditions at each site is different, generally a wave equation-based computer program is required to generate the pile-driving criteria for each individual project. The pile-driving criteria include:

  (i)   hammer stroke versus blow per foot (BPF) (1/set) for required bearing capacity,
 (ii)   maximum compressive stresses versus BPF,
(iii)   maximum tension stress versus BPF.

   However, this process can be rather time consuming and requires very specialized knowledge of the wave equation program.
   The essence of modeling/numerical mapping is prediction, which is obtained by relating a set of variables in input space to a set of response variables in output space through a model. The analysis of pile drivability involves a large number of design variables and nonlinear responses, particularly with statistically dependent inputs. This chapter explores the use of MARS to capture the intrinsic nonlinear and multidimensional relationship associated with pile drivability with a big data set, and a database of more than four thousand piles is utilized for model development and comparative performance between BPNN and MARS predictions.

9.2   The Database and Performance Measures

Table 9.1 shows the performance measures utilized for prediction comparison of the two metahueristic methods.

The database containing 4072 piles with a total of seventeen variables is developed from the information on piles already installed for bridges in the State of North Carolina (Jeon and Rahman 2008). For the whole data sets and the details of each design variables and responses, the report by Joen and Rahman (2008) can be referred to. Seventeen variables including hammer characteristics, hammer cushion material, pile and soil parameters, ultimate pile capacities, and stroke were regarded as inputs to estimate the three dependent responses comprising of the maximum compressive stresses (MCS), maximum tensile stresses (MTS), and blow per foot (BPF). A summary of the input variables and outputs is listed in Table 9.2.

For purpose of simplifying the analyses considering the extensive number of parameters and large data set, Joen and Rahman (2008) divided the data into five categories based on the ultimate pile capacity, as detailed in Table 9.3. In this chapter, for each category 70% of the data patterns were randomly selected as the training data set and the remaining data were used for testing.

Table 9.1 Summary of performance measures

Measure	Calculation
Coefficient of determination (R^2)	$R^2 = 1 - \dfrac{\frac{1}{n}\sum_{i=1}^{n}(Y_i - \overline{Y})^2}{\frac{1}{n}\sum_{i=1}^{n}(y_i - \overline{y})^2}$
Coefficient of correlation (r)	$r = \dfrac{\sum_{i=1}^{N}(Y_i - \overline{Y})(y_i - \overline{y})}{\sqrt{\sum_{i=1}^{N}(Y_i - \overline{Y})^2}\sqrt{\sum_{i=1}^{N}(y_i - \overline{y})^2}}$
Relative root mean squared error (RRMSE)	$RRMSE = \dfrac{\sqrt{\frac{1}{N}\sum_{i=1}^{N}(Y_i - y_i)^2}}{\frac{1}{N}\sum_{i=1}^{N}y_i} \times 100$
Performance index (ρ)	$\rho = \dfrac{RRMSE}{1+r}$

$\overline{y}$ is the mean of the target values of y_i; $\overline{Y}$ is the mean of the predicted Y_i; N denotes the number of data points in the used set, training set, testing set, or the overall set
Definitions of $RRMSE$, r, and ρ are based on Gandomi and Roke (2013)

Figure 9.2 plots the histograms of the seventeen variables.

Table 9.2 Summary of input variables and outputs

Inputs and outputs	Parameters and parameter descriptions		
Input variables	Hammer	Hammer weight (kN)	Variable 1 ($x1$)
		Energy (kN m)	Variable 2 ($x2$)
	Hammer cushion material	Area (m^2)	Variable 3 ($x3$)
		Elastic modulus (GPa)	Variable 4 ($x4$)
		Thickness (m)	Variable 5 ($x5$)
		Helmet weight (kN)	Variable 6 ($x6$)
	Pile information	Length (m)	Variable 7 ($x7$)
		Penetration (m)	Variable 8 ($x8$)
		Diameter (m)	Variable 9 ($x9$)
		Section area (m^2)	Variable 10 ($x10$)
		L/D	Variable 11 ($x11$)
	Soil information	Quake at toe (m)	Variable 12 ($x12$)
		Damping at shaft (s/m)	Variable 13 ($x13$)
		Damping at toe (s/m)	Variable 14 ($x14$)
		Shaft resistance (%)	Variable 15 ($x15$)
	Ultimate pile capacity Q_u (kN)		Variable 16 ($x16$)
	Stroke (m)		Variable 17 ($x17$)
Outputs	Maximum compressive stress MCS (MPa)		
	Maximum tensile stress MTS (MPa)		
	BPF		

Table 9.3 Division of data with respect to ultimate pile capacities

Pile type	Q_u range (kN)	Data		
		No. of training data	No. of testing data	Total
Q_1	133.4–355.9	270	90	360
Q_2	360.0–707.3	428	144	572
Q_3	707.4–1112.1	808	249	1057
Q_4	1112.2–1774.8	1296	421	1717
Q_5	1774.9–3113.7	276	90	366

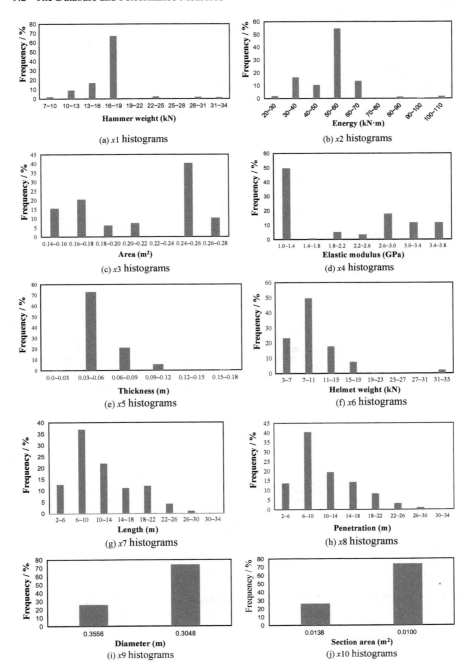

Fig. 9.2 Histograms of the seventeen input variables

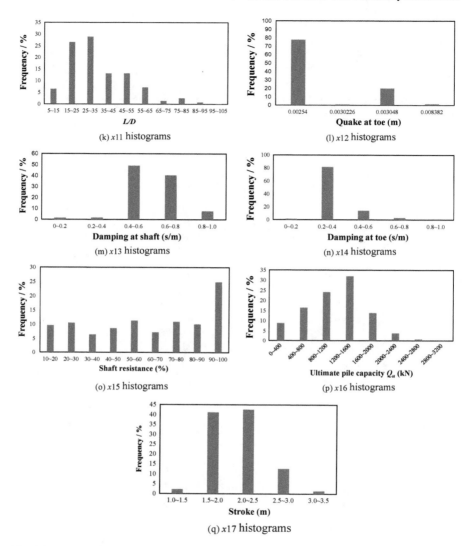

Fig. 9.2   (continued)

9.3   The Developed MARS Models and Modeling Results

9.3.1   Pile Category Q_1

Table 9.4 lists some sample training (Tr.) and testing (Te.) data sets for category Q_1 piles. Q_1 piles have been reanalyzed using MARS and BPNN. For the developed BPNN models, the optimum numbers of hidden neurons are 8, 7, and 9 for MCS,

MTS, and BPF, respectively. The MARS models predicting MCS, MTS, and BPF comprise 25, 43, and 40 BFs, respectively.

A plot of the BPNN and MARS predicted MCS, MTS, and BPF values versus the measured values for the training and testing patterns of Q1 type are shown in Fig. 9.3. Comparison between BPNN and MARS shows that the BPNN model is only slightly more accurate than the MARS model for the training patterns. For the testing results, the MARS model performs slightly better than the BPNN model. Therefore, both MARS and BPNN can serve as reliable tools for the prediction of the pile drivability.

In addition, comparisons of r, RRMSE and ρ in Rows 4, 5, and 6 of Table 9.5 indicate that the differences in the accuracy of the BPNN and MARS models are marginal. Therefore, both methods serve as reliable tools for prediction of pile drivability for Q_1 piles. Table 9.5 also shows the CPU processing time using BPNN and MARS in columns 3 and 10, from which it is obvious that the processing speed of MARS algorithm is much greater, implying that the distinct advantage of MARS over BPNN lies in its convergence speed.

Table 9.6 displays the ANOVA decomposition of the built MARS models for MCS, MTS, and BPF, respectively. For each model, the ANOVA functions are listed. The GCV column provides an indication on the significance of the corresponding ANOVA function, by listing the GCV value for a model with all BFs corresponding to that particular ANOVA function removed. It is this GCV score that is used to assess whether the ANOVA function is making a significant contribution to the model, or whether it just marginally improves the global GCV score. The #basis column gives the number of BFs comprising the ANOVA function and the variable(s) column lists the input variables associated with this ANOVA function.

Table 9.7 lists the BFs of the MCS model and the corresponding equations. The MARS model to estimate MCS for Q_1 is given by

$$
\begin{aligned}
\text{MCS(MPa)} = {} & 100.1 - 44.81 \times \text{BF1} - 1.679 \times \text{BF2} + 8.58 \times \text{BF3} - 7.11 \times \text{BF4} \\
& - 79.87 \times \text{BF5} + 412 \times \text{BF6} - 12.7 \times \text{BF7} + 2.25 \times \text{BF8} + 134 \times \text{BF9} \\
& - 387 \times \text{BF10} + 8.4 \times \text{BF11} - 1817 \times \text{BF12} + 2.58 \times 10^4 \times \text{BF13} \\
& + 0.067 \times \text{BF14} + 85 \times \text{BF15} - 138.7 \times \text{BF16} - 98.1 \times \text{BF17} \\
& - 424.6 \times \text{BF18} + 902 \times \text{BF19} + 85 \times \text{BF20} - 5.7 \times \text{BF21} \\
& - 0.66 \times \text{BF22} - 2.4 \times \text{BF23} + 0.65 \times \text{BF24} - 18.9 \times \text{BF25}
\end{aligned}
\tag{9.1}
$$

Figure 9.4a–d plots the knot locations for $x1$ (Hammer weight), $x6$ (Helmet weight), $x8$ (pile penetration) and $x17$ (stroke), respectively.

Table 9.4 Sample training and testing data sets for category Q_1 type

x1 (kN)	x2 (kN m)	x3 (m²)	x4 (GPa)	x5 (m)	x6 (kN)	x7 (m)	x8 (m)	x9 (m)	x10 (m²)	x11	x12 (m)	x13 (s/m)	x14 (s/m)	x15 (%)	x16 (kN)	x17 (m)	MCS (MPa)	MTS (MPa)	BPF
Training data																			
12.9	34.5	0.18	3.35	0.09	9.37	10.67	9.75	0.36	0.014	27	0.0030	0.66	0.33	95	138	1.26	78.7	11.7	5.5
17.8	43.4	0.26	1.21	0.05	12.54	12.19	10.36	0.30	0.010	34	0.0025	0.59	0.49	95	178	1.18	89.2	6.1	5.3
14.7	36.7	0.15	2.81	0.11	5.83	15.24	15.24	0.30	0.010	50	0.0025	0.59	0.39	85	178	1.19	114.5	12.3	6.3
18.6	58.1	0.26	1.21	0.05	5.19	10.67	10.67	0.30	0.010	35	0.0025	0.72	0.33	75	200	1.17	123.5	2.3	6.4
7.8	27.3	0.20	1.21	0.05	12.41	18.29	18.29	0.30	0.010	60	0.0025	0.59	0.59	92	200	1.95	102.2	8.8	11.1
13.3	35.3	0.26	1.21	0.05	9.46	9.14	9.14	0.30	0.010	30	0.0025	0.66	0.33	15	222	1.70	87.5	0.0	6.6
18.6	58.1	0.16	1.21	0.05	4.00	4.57	4.57	0.30	0.010	15	0.0025	0.59	0.33	45	222	1.05	111.0	0.0	7.3
31.1	81.3	0.26	1.21	0.05	9.52	5.50	5.50	0.30	0.010	18	0.0025	0.50	0.33	57	225	1.22	60.7	0.0	3.0
29.4	102.3	0.27	1.93	0.05	4.00	8.50	8.50	0.30	0.010	28	0.0025	0.66	0.33	87	265	1.23	138.7	0.0	5.1
Testing data																			
12.9	34.5	0.18	3.35	0.09	9.37	10.67	9.75	0.36	0.014	27	0.0030	0.66	0.33	95	138	1.68	112.1	23.8	4.7
14.7	36.7	0.15	2.81	0.11	5.83	15.23	15.23	0.30	0.010	50	0.0025	0.59	0.39	85	178	1.68	154.5	29.4	5.3
18.6	58.1	0.26	1.21	0.05	5.19	12.19	6.40	0.36	0.014	18	0.0031	0.49	0.33	80	200	1.52	142.9	42.3	4.4

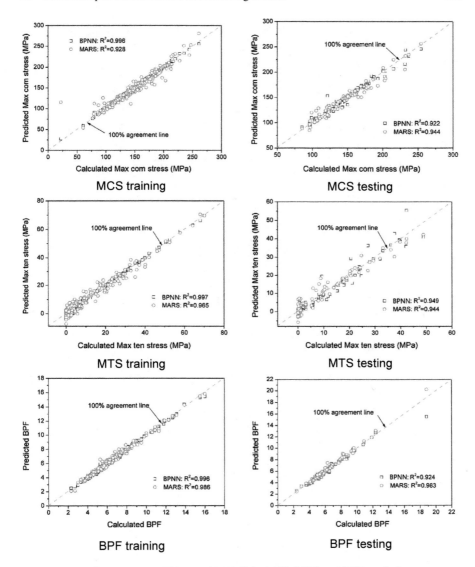

Fig. 9.3 Comparison between BPNN and MARS for MCS, MTS, and BPF predictions

Table 9.5 Comparison of performance measures for BPNN and MARS

Data sets		BPNN								MARS							
		Processing speed (s)	R		RRMSE (%)		ρ (%)			Processing speed (s)	R		RRMSE (%)		ρ (%)		
			Tr.	Te.	Tr.	Te.	Tr.	Te.			Tr.	Te.	Tr.	Te.	Tr.	Te.	
Q_1	MCS	224.88	0.998	0.988	1.508	3.825	0.755	1.924		6.16	0.977	0.916	5.172	9.943	2.616	5.189
	MTS	56.23	0.996	0.984	11.98	22.58	6.001	11.38		9.03	0.983	0.972	25.89	28.99	13.06	14.70
	BPF	200.38	0.999	0.987	0.741	6.361	0.371	3.202		8.82	0.993	0.994	3.949	4.795	1.981	2.405
Q_2	MCS	79.71	0.999	0.996	0.935	1.689	0.468	0.846		18.87	0.997	0.994	1.398	1.949	0.700	0.977
	MTS	208.37	0.998	0.975	11.96	32.54	5.986	16.47		4.54	0.953	0.901	52.04	66.08	26.65	34.76
	BPF	46.26	0.991	0.995	5.691	4.495	2.859	2.253		13.90	0.984	0.989	7.564	6.141	3.813	3.088
Q_3	MCS	285.73	0.991	0.982	2.310	3.106	1.160	1.567		13.74	0.968	0.970	4.180	4.076	2.123	2.069
	MTS	70.10	0.973	0.915	34.53	62.30	17.50	32.53		22.84	0.936	0.869	52.13	73.08	26.93	39.10
	BPF	106.02	0.990	0.975	7.051	9.673	3.543	4.897		33.46	0.972	0.943	11.86	14.87	6.015	7.653
Q_4	MCS	139.40	0.984	0.981	3.064	3.227	1.545	1.629		80.51	0.981	0.979	3.309	3.377	1.670	1.706
	MTS	144.50	0.991	0.876	23.05	65.36	11.58	34.84		83.56	0.989	0.823	25.20	83.14	12.67	45.60
	BPF	137.99	0.962	0.946	13.99	16.93	7.128	8.703		121.30	0.944	0.931	16.90	18.98	8.692	9.827
Q_5	MCS	173.12	0.999	0.997	0.702	1.684	0.351	0.843		10.30	0.997	0.994	1.682	2.136	0.842	1.071
	MTS	148.19	0.999	0.956	4.662	33.71	2.332	17.23		5.22	0.966	0.943	28.11	37.10	14.30	19.09
	BPF	203.11	0.996	0.965	3.534	10.29	1.770	5.238		16.72	0.977	0.924	8.811	14.66	4.456	7.619
Combined Q_1–Q_5	MCS	203.51	0.985	0.987	4.238	3.762	2.135	1.894		131.35	0.978	0.978	5.076	4.810	2.566	2.432
	MTS	363.98	0.893	0.921	80.82	62.36	42.68	32.46		33.11	0.885	0.894	83.60	73.22	44.34	38.65
	BPF	182.68	0.974	0.976	18.48	18.54	9.362	9.388		63.96	0.953	0.960	24.73	23.46	12.67	11.97

Table 9.6 ANOVA decomposition of MARS model for MCS, MTS, and BPF of Q_1

Function	MCS			MTS			BPF		
	GCV	#Basis	Variable(s)	GCV	#Basis	Variable(s)	GCV	#Basis	Variable(s)
1	28.82	1	1	1.047	2	5	39.657	2	1
2	8.346	2	6	575.191	1	6	9.750	2	2
3	7.073	1	8	109.688	2	7	1.760	2	13
4	10.226	1	12	305.352	1	8	3.005	2	15
5	5.629	3	17	251.585	2	11	8.034	2	16
6	11.184	1	13	25.373	1	17	2.976	2	17
7	48.344	2	117	0.441	1	16	66.894	3	13
8	8.048	5	24	337.341	2	37	0.370	2	16
9	11.846	2	34	0.893	2	317	0.235	2	113
10	21.733	2	317	5.626	2	57	0.231	1	116
11	63.062	1	415	2.229	1	511	43.396	2	23
12	8.017	1	68	795.122	4	67	0.357	1	24
13	4.976	3	817	92.069	3	68	0.403	2	216
14				6.797	1	69	0.557	4	217
15				48.170	4	611	0.280	2	313
16				1.472	1	616	0.705	2	415
17				2.593	2	617	0.227	1	417
18				0.626	1	78	0.170	1	513
19				11.173	1	717	0.191	1	615
20				0.447	1	816	0.221	2	715
21				50.089	2	817	0.375	1	1315
22				0.828	1	1115	0.984	1	1617
23				148.475	2	1117			
24				1.472	2	1417			
25				0.466	1	1517			

9.3.2 Pile Categories Q_2–Q_5

The results for the remaining four pile categories Q_2–Q_5 analyzed using MARS and BPNN are described in this section. Table 9.8 lists the number of BFs for MARS model and the number of hidden neurons for BPNN model for each category, respectively.

Comparisons of r, RRMSE, and ρ in Table 9.5 between MARS and BPNN are shown in Rows 7–9, Rows 10–12, Rows 13–15, and Rows 16–18 are for Q_2–Q_5 piles, respectively, from which it is obvious that BPNN gives only slightly more accurate predictions than MARS. Both MARS and BPNN can serve as reliable tools for pile drivability prediction for Q_2–Q_5. Table 9.5 also lists the CPU processing time using BPNN and MARS models for Q_2–Q_5, respectively. The advantage of

Table 9.7   Basis functions and corresponding equations of MARS model for MCS of Q_1

BF	Equation	BF	Equation
BF1	$\max(0, x17 - 2.44)$	BF14	$\max(0, 3.24 - x4) \times \max(0, x15 - 15)$
BF2	$\max(0, x6 - 9.34)$	BF15	$\max(0, x17 - 1.44)$
BF3	$\max(0, 9.34 - x6)$	BF16	$\max(0, 1.44 - x17)$
BF4	$\max(0, 17.8 - x1)$	BF17	$\max(0, 2.44 - x17) \times \max(0, x3 - 0.18)$
BF5	$\max(0, 4.57 - x8)$	BF18	$\max(0, 2.44 - x17) \times \max(0, 0.18 - x3)$
BF6	$\text{BF5} \times \max(0, 2.29 - x17)$	BF19	$\max(0, 3.24 - x4) \times \max(0, x3 - 0.26)$
BF7	$\max(0, 2.44 - x17) \times \max(0, x1 - 29.4)$	BF20	$\max(0, 3.24 - x4) \times \max(0, 0.26 - x3)$
BF8	$\max(0, 2.44 - x17) \times \max(0, 29.4 - x1)$	BF21	$\max(0, 3.24 - x4) \times \max(0, x2\text{-}57.0)$
BF9	$\text{BF5} \times \max(0, x17 - 2.13)$	BF22	$\max(0, 3.24 - x4) \times \max(0, 57.0 - x2)$
BF10	$\text{BF5} \times \max(0, 2.13 - x17)$	BF23	$\max(0, 3.24 - x4) \times \max(0, x2\text{-}43.4)$
BF11	$\max(0, 3.24 - x4) \times \max(0, x2\text{-}54.2)$	BF24	$\max(0, 3.24 - x4) \times \max(0, 43.4 - x2)$
BF12	$\max(0, x1 - 17.8) \times \max(0, 0.18 - x3)$	BF25	$\text{BF5} \times \max(0, 5.83 - x6)$
BF13	$\max(0, 0.0030 - x12)$		

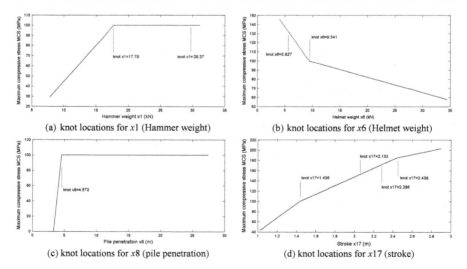

(a) knot locations for $x1$ (Hammer weight)

(b) knot locations for $x6$ (Helmet weight)

(c) knot locations for $x8$ (pile penetration)

(d) knot locations for $x17$ (stroke)

Fig. 9.4   Knot locations for MARS MCS model

No. of	Models	Data			
		Q_2	Q_3	Q_4	Q_5
BFs	MCS	46	30	62	42
	MTS	33	25	65	28
	BPF	39	50	68	50
hidden neurons	MCS	9	10	9	9
	MTS	9	10	10	9
	BPF	8	10	9	9

Table 9.8 No. of hidden neurons/no. of BFs for modeling

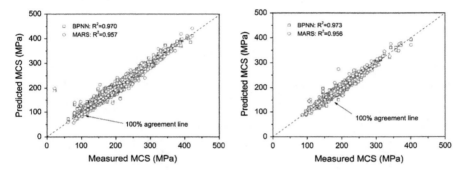

Fig. 9.5 BPNN and MARS estimations versus measured values for MCS model

the processing speed of MARS is obvious, indicating the advantage of MARS over BPNN in computational efficiency. For brevity, the comparison of parameter relative importance and the interpretable MARS models is not elaborated here.

9.4 Combined Data set Q_1–Q_5

Additional analyses were also carried out using the entire 4072 pile data set. For both BPNN and MARS methods, the developed model with the highest R^2 value for the testing data sets is considered to be the optimal. The optimal MARS models to predict MCS, MTS, and BPF use 52, 37, and 35 BFs, respectively. As for the BPNN models, the optimal numbers of hidden neurons are 9, 9, and 8 for MCS, MTS, and BPF, respectively.

Figures 9.5, 9.6, and 9.7 plot the BPNN and MARS estimations versus the measured values for MCS, MTS, and BPF models. For MCS prediction, high R^2 are obtained from both methods. Compared with the MCS predictions, the developed BPNN and MARS models are less accurate in predicting MTS mainly as a result of the bias (errors) due to the significantly smaller tensile stress values. In addition, both BPNN and MARS models can serve as reliable tools for prediction of BPF.

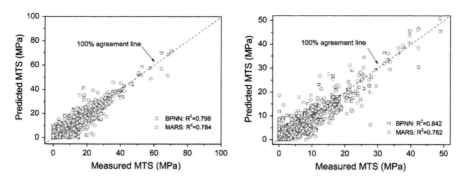

Fig. 9.6 BPNN and MARS estimations versus measured values for MTS model

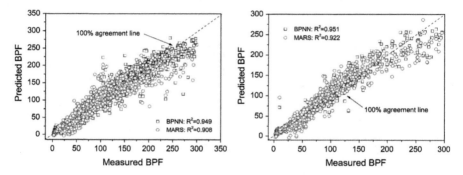

Fig. 9.7 BPNN and MARS estimations versus measured values for BPF model

Comparisons of R^2, r, RRMSE, and ρ in Rows 19–21 of Table 9.5 indicate that BPNN gives only slightly more accurate predictions than MARS. Table 9.5 columns 3 and 10 compare the the CPU processing time (using a PC with 3.0 GHz Intel Core2Quad Q9650 processor, 4 GB RAM). It is obvious that MARS performs better than BPNN in the convergence speed.

9.4.1 Model Interpretability

Table 9.9 lists the BFs of the MCS model and the corresponding equations. The MARS model is in the form of

$$
\begin{aligned}
\text{MCS (MPa)} = {}& 169.4 + 0.0095 \times \text{BF1} + 35.6 \times \text{BF2} - 47.5 \times \text{BF3} - 0.46 \times \text{BF4} \\
& - 2 \times \text{BF5} + 8847 \times \text{BF6} + 9.2 \times \text{BF7} - 8.2 \times \text{BF8} - 0.0025 \times \text{BF9} + 0.0062 \times \text{BF10} \\
& - 3.2 \times \text{BF11} + 470 \times \text{BF12} - 0.0036 \times \text{BF13} - 0.8 \times \text{BF14} - 0.0012 \times \text{BF15} \\
& + 0.006 \times \text{BF16} + 9.43 \times \text{BF17} - 6.1 \times \text{BF18} + 0.136 \times \text{BF19} - 0.098 \times \text{BF20} \\
& - 0.83 \times \text{BF21} - 0.17 \times \text{BF22} - 540 \times \text{BF23} + 1.34 \times 10^5 \text{BF24} + 1.672 \times \text{BF25}
\end{aligned}
$$

$$- 0.42 \times BF26 + 0.144 \times BF27 - 4.57 \times BF28 + 0.0054 \times BF29 + 0.052 \times BF30$$
$$+ 87 \times BF31 + 250 \times BF32 - 763 \times BF33 - 16 \times BF34 - 28.1 \times BF35$$
$$+ 0.217 \times BF36 - 0.2 \times BF37 + 34.5 \times BF38 + 31.3 \times BF39 - 50.2 \times BF40$$
$$- 425 \times BF41 + 0.0018 \times BF42 - 0.003 \times BF43 - 7.4 \times BF44 + 341 \times BF45$$
$$+ 51.4 \times BF46 + 5.67 \times BF47 + 12 \times BF48 + 0.96 \times BF49 + 100.2 \times BF50$$
$$- 0.2 \times BF51 + 0.23 \times BF52. \tag{9.2}$$

Table 9.9 Basis functions and corresponding equations of MARS model for MCS overall data sets

BF	Equation	BF	Equation
BF1	$\max(0, x16 - 1550)$	BF27	$\max(0, x15 - 15)$
BF2	$\max(0, x17 - 2.29)$	BF28	$\max(0, 15 - x15)$
BF3	$\max(0, 2.29 - x17)$	BF29	$BF28 \times \max(0, x16 - 289)$
BF4	$\max(0, x6 - 7.38)$	BF30	$BF28 \times \max(0, 289 - x16)$
BF5	$\max(0, 7.38 - x6)$	BF31	$BF2 \times \max(0, x1 - 29.4)$
BF6	$\max(0, 0.014 - x10)$	BF32	$BF6 \times \max(0, x6 - 6.67)$
BF7	$\max(0, x2 - 30.7)$	BF33	$BF6 \times \max(0, 6.67 - x6)$
BF8	$\max(0, 30.7 - x2)$	BF34	$BF5 \times \max(0, 1.81 - x17)$
BF9	$BF1 \times \max(0, x7 - 8.00)$	BF35	$BF3 \times \max(0, x1 - 29.4)$
BF10	$BF1 \times \max(0, 8.00 - x7)$	BF36	$BF7 \times \max(0, x11 - 50)$
BF11	$\max(0, x11 - 9)$	BF37	$BF7 \times \max(0, 50 - x11)$
BF12	$\max(0, 9 - x11)$	BF38	$BF28 \times \max(0, x13 - 0.59)$
BF13	$\max(0, 1550 - x16) \times \max(0, x8 - 3.05)$	BF39	$BF28 \times \max(0, 0.59 - x13)$
BF14	$\max(0, 1550 - x16) \times \max(0, 3.05 - x8)$	BF40	$BF4 \times \max(0, x5 - 0.05)$
BF15	$\max(0, 1550 - x16) \times \max(0, x6 - 9.34)$	BF41	$BF4 \times \max(0, 0.05 - x5)$
BF16	$\max(0, 1550 - x16) \times \max(0, 9.34 - x6)$	BF42	$\max(0, 1550 - x16) \times \max(0, x11 - 24)$
BF17	$BF6 \times \max(0, x16 - 1067.5)$	BF43	$\max(0, 1550 - x16) \times \max(0, 24 - x11)$
BF18	$BF6 \times \max(0, 1068 - x16)$	BF44	$BF7 \times \max(0, 0.18 - x3)$
BF19	$BF11 \times \max(0, x4 - 3.24)$	BF45	$\max(0, x3 - 0.26)$
BF20	$BF11 \times \max(0, 3.24 - x4)$	BF46	$\max(0, 0.26 - x3)$
BF21	$BF11 \times \max(0, x1 - 29.4)$	BF47	$BF5 \times \max(0, x4 - 1.97)$
BF22	$BF11 \times \max(0, 29.4 - x1)$	BF48	$BF5 \times \max(0, 1.97 - x4)$
BF23	$BF6 \times \max(0, x7 - 3.05)$	BF49	$BF5 \times \max(0, 44.7 - x2)$
BF24	$BF6 \times \max(0, 3.05 - x7)$	BF50	$BF45 \times \max(0, 30 - x11)$
BF25	$BF7 \times \max(0, x17 - 2.90)$	BF51	$BF11 \times \max(0, x2 - 54.2)$
BF26	$BF7 \times \max(0, 2.90 - x17)$	BF52	$BF11 \times \max(0, 54.2 - x2)$

9.4.2   Parameter Relative Importance

It should be pointed out that the parameter relative importance can also be assessed. For MARS, this is carried out by evaluating the GCV increase caused by removing the considered variables from the developed MARS model. For the BPNN, this is commonly carried out using the method by Garson (1991) and discussed by Das and Basudhar (2006). For brevity, the procedures for parametric relative importance of BPNN model have been omitted. Table 9.10 lists the ANOVA decomposition of the built MARS models for MCS, MTS and BPF, respectively. For each model, the ANOVA functions are listed. The GCV column provides an indication on the significance of the corresponding ANOVA function, by listing the GCV value for a model with all BFs corresponding to that particular ANOVA function removed. It is this GCV score that is used to assess whether the ANOVA function is making

Table 9.10   ANOVA decomposition of MARS model for MCS, MTS, and BPF

Function	MCS			MTS			BPF		
	GCV	#Basis	Variable(s)	GCV	#Basis	Variable(s)	GCV	#Basis	Variable(s)
1	28.82	1	1	1.047	2	5	39.657	2	1
2	8.346	2	6	575.191	1	6	9.750	2	2
3	7.073	1	8	109.688	2	7	1.760	2	13
4	10.226	1	12	305.352	1	8	3.005	2	15
5	5.629	3	17	251.585	2	11	8.034	2	16
6	11.184	1	13	25.373	1	17	2.976	2	17
7	48.344	2	117	0.441	1	16	66.894	3	13
8	8.048	5	24	337.341	2	37	0.370	2	16
9	11.846	2	34	0.893	2	317	0.235	2	113
10	21.733	2	317	5.626	2	57	0.231	1	116
11	63.062	1	415	2.229	1	511	43.396	2	23
12	8.017	1	68	795.122	4	67	0.357	1	24
13	4.976	3	817	92.069	3	68	0.403	2	216
14				6.797	1	69	0.557	4	217
15				48.170	4	611	0.280	2	313
16				1.472	1	616	0.705	2	415
17				2.593	2	617	0.227	1	417
18				0.626	1	78	0.170	1	513
19				11.173	1	717	0.191	1	615
20				0.447	1	816	0.221	2	715
21				50.089	2	817	0.375	1	1315
22				0.828	1	1115	0.984	1	16 17
23				148.475	2	1117			
24				1.472	2	1417			
25				0.466	1	1517			

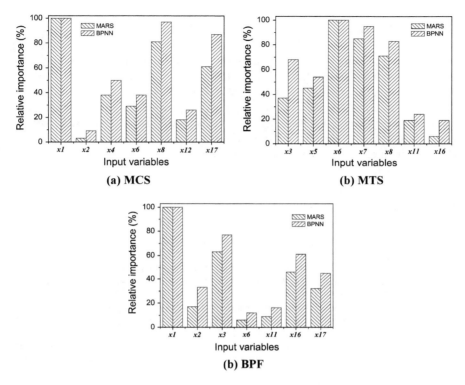

Fig. 9.8 Relative importance of the input variables for HP drivability models developed by MARS and BPNN

a significant contribution to the model, or whether it just marginally improves the global GCV score. The #basis column gives the number of BFs comprising the ANOVA function, and the variable(s) column lists the input variables associated with this ANOVA function.

Figure 9.8 gives the plot of the relative importance of the input variables for the three HP drivability models developed by MARS and BPNN, for comparison. It can be observed that both MCS and BPF are mostly influenced by $x1$ (hammer weight). Interestingly, MTS is primarily influenced by $x6$ (the weight of helmet). It should be noted that since the BPNN and MARS algorithms adopt different methods in assessing the parametric relative importance, it is understandable that two algorithms give different results.

9.5   Summary

This chapter examines the MARS use for multivariate geotechnical problems with a big data set. The chosen database consisted of 4072 HP-pile test results from 67 projects, with a variety of 17 input variables. Both BPNN and MARS models are developed for assessing pile drivability in relation to the prediction of the maximum compressive stresses, maximum tensile stresses, and blow per foot. Comparison with regard to the computational time indicates that the MARS use in multivariate problems with big data sets is more advantageous than BPNN.

References

Das SK, Basudhar BK (2006) Undrained lateral load capacity of piles in clay using artificial neural networks. Comput Geotech 33(8):454–459
Gandomi AH, Roke DA (2013) Intelligent formulation of structural engineering systems. In: Seventh MIT conference on computational fluid and solid mechanics-focus: multiphysics & multiscale, 12–14 June, Cambridge, USA
Garson GD (1991) Interpreting neural-network connection weights. AI Expert 6(7):47–51
Jeon JK, Rahman MS (2008) Fuzzy neural network models for geotechnical problems. Research Project FHWA/NC/2006–52. North Carolina State University, Raleigh, NC
Smith EAL (1960) Pile driving analysis by the wave equation. J Eng Mech Division ASCE 86:35–61

Chapter 10
MARS Use in Estimation of Liquefaction-Induced Lateral Spreading

Soil liquefaction during earthquakes can result in ground movements that cause damage to buildings and lifelines. Lateral spreading is one form of earthquake-induced ground movements that have caused extensive damage in previous earthquakes. The lateral displacement is dependent on many factors including the earthquake magnitude, thickness and particle size of the liquefiable subsoils, and the depth of the groundwater. A number of analytical and empirical methods have been proposed to predict the magnitude of the lateral displacement. One common empirical method which is called MLR model is based on multiple linear regression (MLR) analysis of a database of observed case histories. It is proposed in this chapter to use MARS to predict the liquefaction-induced lateral displacement.

10.1 Background

During an earthquake, liquefaction occurs in saturated sand deposits, due to excess pore water pressure increase. It can cause serious to destructive damages to structures. The liquefaction mechanism includes ground subsidence, flow failure, and lateral spreading, among other effects. Lateral spreading involves the movement of relatively intact soil blocks on a layer of liquefied soil toward a free face or down a gentle slope. It can also induce different forms of ground deformation, the magnitudes of which range from a few centimeters to several meters. Susceptibility to liquefaction-induced lateral spreading is dependent on a number of factors such as the depth of the groundwater table, the physical and mechanical properties of the subsoils, and the intensity and duration of the ground shaking. The large number of factors involved presents challenges in developing simplified analytical solutions to estimate the magnitude of the lateral displacement. A rigorous numerical model must consider dynamic and three-dimensional effects as well as the anisotropic and heterogeneous nature of liquefiable soil deposits. Moreover, accurate constitutive modeling of a liquefiable soil is a difficult problem, even when considerable laboratory testing is undertaken. Such efforts are hampered by the difficulty in obtaining representative,

© Science Press and Springer Nature Singapore Pte Ltd. 2019
W. Zhang, *MARS Applications in Geotechnical Engineering Systems*,
https://doi.org/10.1007/978-981-13-7422-7_10

"undisturbed" testing samples from the in situ deposit. In practice, empirical models based on case histories have remained the more popular method in past decades.

Various empirical approaches have also been proposed (e.g., Hamada et al. 1986; Youd and Perkins 1987; Bartlett and Youd 1992a, b; 1995; Shamoto et al. 1998; Rauch and Martin 2000; Bardet et al. 1999, 2002; Youd et al. 2002; Zhang et al. 2004; Zhang and Zhao 2005; Al Bawwab 2005; Kanibir et al. 2006; Aydan et al. 2008). Table 10.1 shows some of the empirical models, in which the physical meanings of the symbols can be referred in Table 10.2.

The most widely used method is the multiple linear regression (MLR) approach originally proposed by Bartlett and Youd (1995) in which two different site conditions are considered: (1) lateral spread toward a free face (e.g., river) and (2) lateral spread down gentle ground slopes where a free face is absent or far away. Based on database records from case histories, empirical models were developed for estimating horizontal ground displacement from liquefaction-induced lateral spread. The original procedure was later revised (Youd et al. 1999). The most updated version of the equations is as follows (Youd et al. 2002):

For free-face conditions:

$$\begin{aligned} \log D_H = {}& -16.713 + 1.532M - 0.012R - 1.406\log(R^*) + 0.592\log(W) \\ & + 0.540\log(T_{15}) + 3.413\log(100 - F_{15}) - 0.795\log(D50_{15} + 0.1) \end{aligned}$$
$$(10.1)$$

and for gently sloping ground:

$$\begin{aligned} \log D_H = {}& -16.213 + 1.532M - 0.012R - 1.406\log(R^*) + 0.338\log(S) \\ & + 0.540\log(T_{15}) + 3.413\log(100 - F_{15}) - 0.795\log(D50_{15} + 0.1) \end{aligned}$$
$$(10.2)$$

in which the parameters and parameter descriptions are listed in Table 10.2. The MLR models of Youd et al. (2002) are the most commonly referred to estimate the liquefaction-induced lateral spread. However, its predictive accuracy is not still satisfactory.

A soft computing technique known as artificial neural networks (ANNs) has also been successfully applied to estimate the lateral displacement based on case records (Wang and Rahman 1999; Chiru-Danzer et al. 2001). The main advantage of ANN over other regression techniques is the ability to capture and represent the nonlinear interaction among the multitude of variables of the system without having to assume the form of the relationship between the variables. Some limitations of neural networks include model interpretability, computational intensity, slow convergence speed, and overfitting problems.

In the light of the above-mentioned techniques or models, the liquefaction-induced lateral spread database used by Youd et al. (2002) has been reanalyzed using MARS. The developed MARS models, the predictive accuracy, model interpretability, parametric sensitivity analysis, as well as some design charts from the built MARS model are presented.

Table 10.1 Some empirical models for prediction of the lateral spread

Method	Subset	Model
Hamada et al. (1986)		$D_H = 0.75 H^{1/2}\theta^{1/3}$
Youd and Perkins (1987)		$\text{Log LSI} = -3.49 - 1.86\,\text{Log}\,R + 0.98 M_w$
Bardet et al. (1999)	ff	$\text{Log}\,(D_H + 0.01) = -17.372 + 1.248 M_w - 0.923\,\text{Log}\,R - 0.014 R$ $+ 0.685\,\text{Log}\,W + 0.3\,\text{Log}\,T_{15} + 4.826\,\text{Log}\,(100 - F_{15}) - 1.091\,D50_{15}$
	gs	$\text{Log}\,(D_H + 0.01) = -14.152 + 0.988 M_w - 1.049\,\text{Log}\,R - 0.011 R$ $+ 0.318\,\text{Log}\,S + 0.619\,\text{Log}\,T_{15} + 4.287\,\text{Log}\,(100 - F_{15}) - 0.705\,D50_{15}$
Youd et al. (2002)	ff	$\text{Log}\,D_H = -16.713 + 1.532 M_w - 1.406\,\text{Log}\,R^* - 0.012 R + 0.592\,\text{Log}\,W$ $+ 0.540\,\text{Log}\,T_{15} + 3.413\,\text{Log}\,(100 - F_{15}) - 0.795\,\text{Log}\,(D50_{15} + 0.1\,\text{mm})$
	gs	$\text{Log}\,D_H = -16.213 + 1.532 M_w - 1.406\,\text{Log}\,R^* - 0.012 R + 0.338\,\text{Log}\,S$ $+ 0.540\,\text{Log}\,T_{15} + 3.413\,\text{Log}\,(100 - F_{15}) - 0.795\,\text{Log}\,(D50_{15} + 0.1\,\text{mm})$

Table 10.2   Parameters and parameter descriptions used in Youd et al. (2002)

Parameters		Parameter description
Output	$\log(D_h)$	D_H, the estimated lateral ground displacement, in meters
Input	M	The moment magnitude of the earthquake
	R	The nearest horizontal distance from the site to the seismic energy source, in kilometers
	$\log(R^*)$	R^*, the modified source distance, in kilometers
	$\log(W)$ or $\log(S)$	$\log(W)$ for free-face condition, W, the free-face ratio defined as the height (H) of the free face divided by the horizontal distance (L) from the base of the free face to the point in question, in percent $\log(S)$ for gentle slope ground, S, the ground slope, in percent
	$\log(T_{15})$	T_{15} is the cumulative thickness of saturated granular layers with corrected blow counts, $(N_1)_{60}$, less than 15, in meters
	$\log(100 - F_{15})$	F_{15} is the average fines content for granular materials included within T_{15}, in percent
	$\log(D50_{15} + 0.1)$	$D50_{15}$ is the average mean grain size for granular materials within T_{15}, in millimeters

10.2   The Database

Bartlett and Youd (1995) compiled case histories of lateral spread for the sites and earthquakes, and in total, 467 displacement vectors from the case history database were used to fit the final MLR model. A complete tabulation of the database is available from the National Center for Earthquake Engineering Research (Bartlett and Youd 1992a). The amount of horizontal ground displacement at the case history sites was measured or estimated by several investigators using various methods. Measurements of the independent variables were obtained from seismological reports, topographical maps, and surface investigations. In Youd et al. (2002), additional case history data were added from three earthquakes—1983 Borah Peak, Idaho, 1989 Loma Prieta, California, and 1995 Hyogo-Ken Nanbu (Kobe), Japan. The added sites contain data from several course-grained liquefiable sites, allowing the extension of the predictive equation to coarser-grained materials than is allowed by Bartlett and Youd (1992a, 1995). Youd listed the compiled case history data, including the newly added data on his Web site. This chapter utilizes this database comprising of 228 vectors related to free-face condition and 256 vectors involving gently sloping ground. These 484 lateral spread case records can be referred to Youd et al. (2002). For the free-face ground, a total of 170 cases were randomly selected as training data and the remaining 58 cases as testing (validation) data. For the gently sloping ground, 192 cases were randomly selected as training data and 64 cases as testing data.

10.3 The Developed MARS Model

The prediction of lateral spread displacement using MARS model with second-order interaction adopted 17 BFs of linear spline function for free-face ground condition and 18 BFs of linear spline function for gentle sloping ground, respectively. The execution time using a PC with 3.0 GHz Intel Core2Quad Q9650 processor (4 GB RAM) is 1.94 s for free-face ground condition and 1.08 s for gently sloping ground. A comparison of lateral spread displacement predictions for MLR and MARS is shown in Figs. 10.1 and 10.2. The plots of the predicted versus measured displacements in Figs. 10.1 and 10.2 as well as the results in terms of R^2 and the root mean square error (RMSE) as summarized in Table 10.3 indicate that the MARS predictions are more accurate than the MLR model. For the MARS training data sets, the R^2 of 0.9368 for free-face condition and R^2 of 0.9047 for gentle slope are much higher than the counterparts of MLR, respectively. The lower RMSE values indicate that the errors in MARS training results are even smaller. As for the testing samples, the R^2 of 0.8559 for free face is slightly higher than that of MLR, while the R^2 of 0.8939 for gentle slope is considerably greater than that of MLR. The RMSE values of the MARS testing patterns are lower than those of MLR. It should be noted that the MLR model was developed using the entire database and it was not validated against an independent testing set, whereas the MARS models were developed using about 75% of the cases with the remaining 25% as the validation set.

The measured and predicted $\log(D_h)$ values for the testing data as summarized in Tables 10.4 and 10.5 indicate that for the majority of the cases, the relative error (relative error is defined as the ratio of the difference between the measured and predicted $\log(D_h)$ value to the measured value, in percentage) is smaller for MARS compared with MLR predictions.

Figure 10.3 presents the histogram plots of the ratio of the predicted to measured displacement values. For free-face condition, more than 50% of the MARS predic-

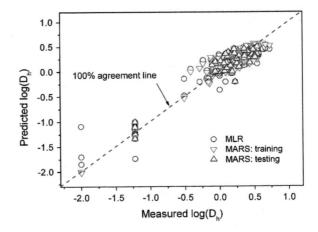

Fig. 10.1 Comparison between MLR and MARS results for free-face ground

Fig. 10.2 Comparison
between MLR and MARS
results for gently sloping
ground

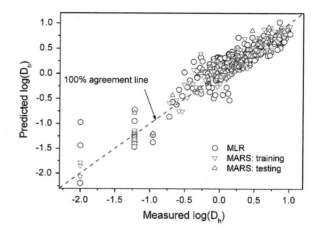

Table 10.3 Comparison
between MARS and MLR
results

Ground conditions	Data set	R^2		RMSE	
		MLR	MARS	MLR	MARS
Free face	Training	0.8358	0.9368	0.2183	0.1449
	Testing		0.8559		0.1530
Gently sloping	Training	0.8342	0.9047	0.1719	0.1285
	Testing		0.8939		0.1429

tions of the data patterns fell within ±20% of the measured values, and more than 80% of the predictions were within ±40% of the measured values. In comparison, using MLR, about 36% of the predictions fell within ±20% of the measured values, and about 66% of the predictions were within ±40% of the measured values. For gently sloping ground, more than 52% of the MARS predictions of the data patterns fell within ±20% of the measured values, and more than 85% of the predictions were within ±40% of the measured values. In comparison, using MLR, about 44% of the predictions fell within ±20% of the measured values, and about 72% of the predictions were within ±40% of the measured values. Figure 10.4 conceptually lists the relative errors of both the free-face and gentle slope case points from both approaches. The results indicate that the MARS model in predicting liquefaction-induced lateral spread is an improvement over the widely used MLR method proposed by Youd et al. (2002).

Figure 10.5 presents the predicted versus measured results for the entire range of lateral displacements. Since it may be difficult to conceive engineering counter-measures in situations involving large displacements, engineers are generally more interested in predictions of moderate displacements of up to about 2 m (Zhang 2001). The lateral displacements of up to 2 m as shown in Fig. 10.6 indicate that the majority of predicted displacements using the MARS model plot within a factor of two of the measured values.

Table 10.4 Measured and predicted log(D_h) as well as the relative error by MLR and MARS for free-face ground

Testing data No.	Earthquakes/References	log(D_h) (Measured)	log(D_h) (MLR)	log(D_h) (MARS)	Relative error (%) MLR	MARS
1	1964, ALASKA	0.2695	0.4422	0.1827	64	32
2	1964, NIIGATA	0.1335	0.3916	0.3690	193	176
3	1964, NIIGATA	0.2695	0.3553	0.3491	32	30
4	1964, NIIGATA	−0.2518	0.2464	0.1061	198	142
5	1964, NIIGATA	0.1173	0.2374	0.2914	102	148
6	1964, NIIGATA	0.4579	0.2914	0.3501	36	24
7	1964, NIIGATA	0.5159	0.5229	0.3919	1	24
8	1964, NIIGATA	0.4639	0.7158	0.8551	54	84
9	1964, NIIGATA	0.6609	0.5298	0.6453	20	2
10	1964, NIIGATA	0.7160	0.3709	0.4538	48	37
11	1964, NIIGATA	0.2601	0.1798	0.1456	31	44
12	1964, NIIGATA	0.3820	0.1873	0.2673	51	30
13	1964, NIIGATA	0.4330	0.2760	0.1951	36	55
14	1964, NIIGATA	0.2923	0.2082	0.2131	29	27
15	1964, NIIGATA	0.9227	0.6331	0.8409	31	9
16	1964, NIIGATA	0.7910	0.6832	0.7652	14	3
17	1964, NIIGATA	−0.1427	−0.3684	−0.0208	158	85
18	1964, NIIGATA	0.1206	−0.3965	−0.0574	429	148
19	1964, NIIGATA	−0.0410	0.2454	0.0200	699	149
20	1964, NIIGATA	−0.1612	−0.2701	−0.1150	68	29
21	1964, NIIGATA	0.6840	0.6432	0.7017	6	3
22	1964, NIIGATA	0.2625	0.3424	0.2769	30	6
23	1964, NIIGATA	0.8488	0.6953	0.7643	18	10
24	1964, NIIGATA	0.3997	0.4832	0.4870	21	22
25	1964, NIIGATA	0.4728	0.4535	0.4498	4	5
26	1964, NIIGATA	0.5453	0.5947	0.5895	9	8
27	1964, NIIGATA	0.3202	0.3179	0.2308	1	28
28	1964, NIIGATA	0.5775	0.6213	0.7152	8	24
29	1964, NIIGATA	0.7016	0.8817	0.9229	26	32
30	1964, NIIGATA	0.1732	0.2698	0.1650	56	5
31	1964, NIIGATA	0.6884	0.7168	0.8148	4	18
32	1964, NIIGATA	0.2856	0.3336	0.2555	17	11
33	1964, NIIGATA	0.0253	0.1834	0.0413	625	63
34	1964, NIIGATA	−0.0410	0.1010	−0.0601	347	47

(continued)

Table 10.4 (continued)

Testing data No.	Earthquakes/References	$\log(D_h)$ (Measured)	$\log(D_h)$ (MLR)	$\log(D_h)$ (MARS)	Relative error (%)	
					MLR	MARS
35	1964, NIIGATA	0.1038	0.2120	0.0618	104	40
36	1964, NIIGATA	0.2148	0.1900	0.0460	12	79
37	1964, NIIGATA	0.3324	0.2674	0.1015	20	69
38	1964, NIIGATA	0.7782	0.6004	0.6212	23	20
39	1964, NIIGATA	0.5289	0.3907	0.3672	26	31
40	1964, NIIGATA	0.0899	0.1343	−0.0885	49	198
41	1964, NIIGATA	0.1644	0.2152	0.0763	31	54
42	1971, SAN FERNANDO	0.4997	0.2707	0.3767	46	25
43	1971, SAN FERNANDO	−0.2840	0.1184	−0.2797	142	2
44	1971, SAN FERNANDO	0.4440	0.3947	0.4769	11	7
45	1971, SAN FERNANDO	0.3054	0.1980	0.2546	35	17
46	1971, SAN FERNANDO	0.3096	0.2737	0.4181	12	35
47	1979, IMPERIAL VALLEY	−0.1739	0.0050	−0.0471	103	73
48	1979, IMPERIAL VALLEY	0.0531	0.0982	0.1357	85	156
49	1979, IMPERIAL VALLEY	0.5821	0.3902	0.5437	33	7
50	1979, IMPERIAL VALLEY	0.3617	0.3475	0.3473	4	4
51	1979, IMPERIAL VALLEY	0.1790	0.2644	0.2147	48	20
52	1979, IMPERIAL VALLEY	−0.1427	0.1332	0.1123	193	179
53	1987, SUPERSTITION HILLS	−0.6778	−0.5681	−0.4294	16	37
54	1987, SUPERSTITION HILLS	−0.9586	−1.2283	−1.1918	28	24
55	Ambraseys (1988)	−1.2219	−0.7232	−0.7830	41	36
56	1995 Hyogo-Ken Nanbu	0.1673	0.1235	0.1750	26	5
57	1995 Hyogo-Ken Nanbu	−0.1805	−0.1526	−0.0562	15	69
58	1995 Hyogo-Ken Nanbu	0.0719	0.3239	0.0347	351	52

Table 10.5 Measured and predicted $\log(D_h)$ as well as the relative error by MLR and MARS for gently sloping ground

Testing data No.	Earthquakes/References	$\log(D_h)$ (Measured)	$\log(D_h)$ (MLR)	$\log(D_h)$ (MARS)	Relative error (%) MLR	Relative error (%) MARS
1	Ambraseys (1988)	−1.2219	−1.2159	−1.3778	0	13
2	Ambraseys (1988)	−1.2219	−1.1053	−1.0571	10	13
3	Ambraseys (1988)	−1.2219	−1.3390	−1.4732	10	21
4	Ambraseys (1988)	−1.2219	−1.2432	−1.2998	2	6
5	Ambraseys (1988)	−1.2219	−1.0037	−0.8837	18	28
6	1971, SAN FERNANDO	0.2279	−0.2051	0.0662	190	71
7	1964, NIIGATA	0.1931	0.3161	0.4174	64	116
8	1964, NIIGATA	0.2305	0.2111	0.1320	8	43
9	1964, NIIGATA	0.1303	0.1877	0.0608	44	53
10	1964, NIIGATA	−0.0506	0.1600	−0.0240	416	53
11	1964, NIIGATA	0.1959	0.2022	0.1048	3	47
12	1964, NIIGATA	0.4409	0.1773	0.0288	60	93
13	1964, NIIGATA	0.2553	0.1660	−0.0058	35	102
14	1964, NIIGATA	0.1303	0.3884	0.4176	198	220
15	1964, NIIGATA	0.4955	0.4491	0.5237	9	6
16	1964, NIIGATA	0.5185	0.4368	0.4971	16	4
17	1964, NIIGATA	0.4654	0.4613	0.5022	1	8
18	1964, NIIGATA	0.4116	0.4070	0.4360	1	6
19	1964, NIIGATA	0.5328	0.4456	0.4558	16	14
20	1964, NIIGATA	0.5441	0.4428	0.5097	19	6
21	1964, NIIGATA	−0.0809	−0.0190	0.0138	77	117
22	1964, NIIGATA	0.1004	0.1304	0.1068	30	6
23	1964, NIIGATA	0.0792	0.0047	0.1544	94	95
24	1964, NIIGATA	0.1818	−0.0378	0.0174	121	90
25	1964, NIIGATA	−0.0177	−0.0075	0.1176	58	763
26	1964, NIIGATA	0.1875	−0.0016	0.0383	101	80
27	1964, NIIGATA	0.5465	0.4963	0.5775	9	6
28	1964, NIIGATA	0.4378	0.2877	0.2667	34	39
29	1964, NIIGATA	0.3692	0.2575	0.2377	30	36
30	1964, NIIGATA	0.5465	0.3299	0.3349	40	39
31	1964, NIIGATA	0.4871	0.3453	0.3535	29	27
32	1964, NIIGATA	0.4742	0.3373	0.3503	29	26
33	1964, NIIGATA	0.4928	0.2894	0.4102	41	17
34	1964, NIIGATA	0.5340	0.3720	0.4735	30	11
35	1964, NIIGATA	0.5465	0.3494	0.4540	36	17
36	1964, NIIGATA	0.5159	0.3197	0.4282	38	17
37	1964, NIIGATA	0.5159	0.2889	0.3595	44	30
38	1964, NIIGATA	0.1790	0.2891	0.2818	62	57
39	1964, NIIGATA	0.1173	0.2659	0.1813	127	55

(continued)

Table 10.5 (continued)

Testing data No.	Earthquakes/References	$\log(D_h)$ (Measured)	$\log(D_h)$ (MLR)	$\log(D_h)$ (MARS)	Relative error (%) MLR	MARS
40	1964, NIIGATA	−0.1024	0.1310	0.1927	228	288
41	1964, NIIGATA	0.1644	0.3282	0.1890	100	15
42	1964, NIIGATA	0.0645	0.2831	0.1835	339	185
43	1964, NIIGATA	0.2856	0.3214	0.2553	13	11
44	1964, NIIGATA	0.1492	0.2837	0.2542	90	70
45	1983, NIHONKAI-CHUBU	0.0940	0.1252	0.0369	33	61
46	1983, NIHONKAI-CHUBU	0.3606	0.2763	0.2064	23	43
47	1983, NIHONKAI-CHUBU	0.3705	0.3474	0.2650	6	28
48	1983, NIHONKAI-CHUBU	0.3606	0.4194	0.3637	16	1
49	1983, NIHONKAI-CHUBU	0.3656	0.3135	0.2617	14	28
50	1983, NIHONKAI-CHUBU	0.0177	0.2090	0.1614	1081	812
51	1983, NIHONKAI-CHUBU	0.3606	0.4565	0.3209	27	11
52	1983, NIHONKAI-CHUBU	0.2222	0.3677	0.2608	65	17
53	1983, NIHONKAI-CHUBU	0.4891	0.4397	0.4329	10	11
54	1983, NIHONKAI-CHUBU	0.1208	0.3045	0.2142	152	77
55	1983, NIHONKAI-CHUBU	0.1120	0.0697	0.0562	38	50
56	1983, NIHONKAI-CHUBU	0.2153	0.4528	0.3056	110	42
57	1983, NIHONKAI-CHUBU	0.0940	0.3531	0.2948	276	214
58	1983, NIHONKAI-CHUBU	0.1031	0.1797	0.1111	74	8
59	1983, NIHONKAI-CHUBU	−0.0563	0.1978	0.1353	451	340
60	1983, NIHONKAI-CHUBU	0.2551	0.2552	0.1802	0	29
61	1983, NIHONKAI-CHUBU	0.1620	0.1216	0.0674	25	58
62	1983, NIHONKAI-CHUBU	0.0694	0.1650	0.1225	138	77
63	1983, NIHONKAI-CHUBU	0.2870	0.0954	0.0338	67	88
64	1983, NIHONKAI-CHUBU	−0.0563	0.0010	−0.0526	102	7

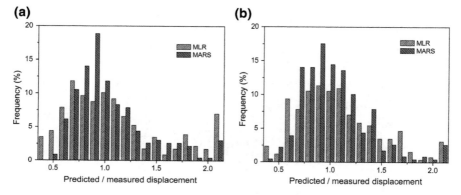

Fig. 10.3 Histogram of predicted/measured displacement using: **a** MLR and MARS for free face, **b** MLR and MARS for gentle slope

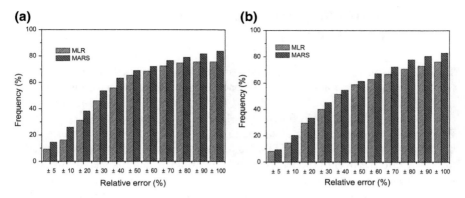

Fig. 10.4 Variation of the relative errors obtained from the MRL and MARS: **a** free face and **b** gentle slope

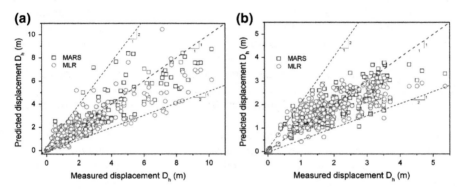

Fig. 10.5 Predicted versus measured displacements using MARS and MLR: **a** free face and **b** gentle slope

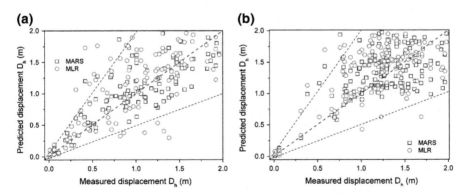

Fig. 10.6 Predicted versus measured displacements up to 2 m using MARS and MLR: **a** free face and **b** gentle slope

10.4   Parametric Sensitivity Analysis and Model Interpretability

Table 10.6 displays the ANOVA decomposition of the developed MARS models for free-face and gently sloping grounds. For each MARS model, the first column lists the ANOVA function number. The second column gives an indication of the importance of the corresponding ANOVA function, by listing the GCV score for a model with all BFs corresponding to that particular ANOVA function removed. The third column provides the standard deviation of this function. It gives an indication of its relative importance to the overall model and can be interpreted in a manner similar to the standardized regression coefficient in a linear model. The fourth column gives the number of BFs comprising the ANOVA function. The last column gives the particular input variables associated with the ANOVA function. Figure 10.7 gives the plots of the relative importance of the input variables for the two MARS models, which is evaluated by the increase in the GCV value caused by removing the considered variables from the developed MARS model. The results indicate that for the free-face condition, the two most important variables are variable 4 [log(W), the logarithmic value of free-face ratio] and variable 2 (R, distance from the site to the seismic energy source). For the gently sloping ground, the two most important variables are variable 1 (M, the moment magnitude of the earthquake) and variable 2.

Table 10.7 lists the BFs and their corresponding equations for the developed MARS models.

It is observed from Table 10.7 that interactions have occurred between BFs. The presence of interactions suggests that these two models are not simply additive and that interactions play an important role in building accurate models for lateral spread displacement prediction. The equation of the optimum MARS model for free-face ground is given by

$$
\begin{aligned}
\log(D_h) = {} & 0.464 - 2.022 \times \text{BF1} + 3.456 \times \text{BF2} + 0.919 \times \text{BF3} - 0.027 \\
& \times \text{BF4} + 0.270 \times \text{BF5} - 6.056 \times \text{BF6} - 1.477 \times \text{BF7} \\
& + 1.555 \times \text{BF8} + 0.867 \times \text{BF9} - 2.300 \times \text{BF10} + 8.488 \\
& \times \text{BF11} + 0.597 \times \text{BF12} + 6.796 \times \text{BF13} - 3.508 \\
& \times \text{BF14} + 0.071 \times \text{BF15} - 0.116 \times \text{BF16} + 16.577 \times \text{BF17} \quad (10.3)
\end{aligned}
$$

The equation of the optimum MARS model for gently sloping ground is

$$
\begin{aligned}
\log(D_h) = {} & -1.766 - 1.647 \times \text{BF1} + 3.102 \times \text{BF2} + 1.78 \times \text{BF3} - 0.035 \\
& \times \text{BF4} + 0.08 \times \text{BF5} + 0.798 \times \text{BF6} - 0.036 \times \text{BF7} - 13.161 \\
& \times \text{BF8} + 0.52 \times \text{BF9} - 0.658 \times \text{BF10} - 3.312 \times \text{BF11} - 0.976 \\
& \times \text{BF12} - 0.662 \times \text{BF13} + 35.986 \times \text{BF14} - 3.357 \\
& \times \text{BF15} + 18.876 \times \text{BF16} - 17.095 \times \text{BF17} + 1.864 \times \text{BF18} \quad (10.4)
\end{aligned}
$$

Table 10.6 ANOVA decomposition for MARS models

MARS model for free-face condition					MARS model for gently sloping ground				
Function	GCV	STD	#basis	Variable(s)	Function	GCV	STD	#basis	Variable(s)
1	0.970	0.451	1	M	1	1.064	0.936	2	M
2	1.057	0.776	2	R	2	0.766	0.559	2	R
3	0.870	0.445	1	$\log(W)$	3	0.159	0.156	2	$\log(T_{15})$
4	1.300	0.144	3	$\log(T_{15})$	4	0.159	0.245	1	$\log(100 - F_{15})$
5	0.486	0.454	1	$\log(100 - F_{15})$	5	0.122	0.208	2	$\log(D50_{15} + 0.1)$
6	0.356	0.258	1	$\log(D50_{15} + 0.1)$	6	0.169	0.291	2	$M, \log(R^*)$
7	0.182	0.272	2	$M, \log(W)$	7	0.039	0.083	1	$M, \log(100 - F_{15})$
8	0.041	0.047	1	$R, \log(T_{15})$	8	0.185	0.307	2	$R, \log(T_{15})$
9	0.087	0.168	1	$R, \log(D50_{15} + 0.1)$	9	0.141	0.251	2	$\log(S), \log(T_{15})$
10	0.083	0.144	2	$\log(W), \log(T_{15})$	10	0.500	0.503	2	$\log(S), \log(100 - F_{15})$
11	0.049	0.084	1	$\log(W), \log(100 - F_{15})$					
12	0.303	0.276	1	$\log(T_{15}), \log(100 - F_{15})$					

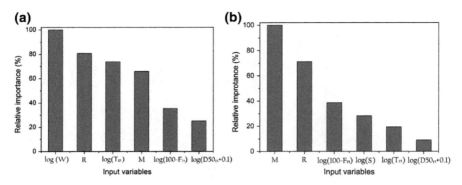

Fig. 10.7 Relative importance of the input variables selected in the MARS models: **a** free face and **b** gentle slope

Table 10.7 Expressions of BFs for MARS models

MARS model for free-face condition		MARS model for gently sloping ground	
BF	Equation	BF	Equation
BF1	$\max(0, \log(T_{15}) - 0.699)$	BF1	$\max(0, 7.5 - M)$
BF2	$\max(0, 0.699 - \log(T_{15}))$	BF2	$BF1 \times \max(0, \log(R^*) - 0.8873)$
BF3	$\max(0, M - 6.6)$	BF3	$BF1 \times \max(0, 0.8873 - \log(R^*))$
BF4	$\max(0, R - 6)$	BF4	$\max(0, R - 27)$
BF5	$\max(0, 6 - R)$	BF5	$\max(0, 27 - R)$
BF6	$\max(0, 1.9315 - \log(100 - F_{15}))$	BF6	$BF5 \times \max(0, \log(T_{15}) - 0.9445)$
BF7	$\max(0, 1.2495 - \log(W))$	BF7	$BF5 \times \max(0, 0.9445 - \log(T_{15}))$
BF8	$\max(0, -0.2441 - \log(D50_{15} + 0.1))$	BF8	$\max(0, \log(S) + 0.6198) \times \max(0, \log(T_{15}) - 1.0294)$
BF9	$BF7 \times \max(0, M - 6.8)$	BF9	$\max(0, \log(S) + 0.6198) \times \max(0, 1.0294 - \log(T_{15}))$
BF10	$BF7 \times \max(0, 6.8 - M)$	BF10	$\max(0, \log(D50_{15} + 0.1) + 0.4763)$
BF11	$BF7 \times \max(0, \log(T_{15}) - 1.1038)$	BF11	$\max(0, -0.4763 - \log(D50_{15} + 0.1))$
BF12	$BF7 \times \max(0, 1.1038 - \log(T_{15}))$	BF12	$\max(0, \log(T_{15}) - 0.7404)$
BF13	$BF6 \times \max(0, 0.906 - \log(W))$	BF13	$\max(0, 0.7404 - \log(T_{15}))$
BF14	$\max(0, 0.9823 - \log(T_{15}))$	BF14	$BF1 \times \max(0, \log(100 - F_{15}) - 1.8808)$
BF15	$BF8 \times \max(0, R - 21)$	BF15	$\max(0, 1.9912 - \log(100 - F_{15}))$
BF16	$BF5 \times \max(0, 0.5682 - \log(T_{15}))$	BF16	$\max(0, \log(S) + 0.6198) \times \max(0, 1.9868 - \times 6)$
BF17	$BF1 \times \max(0, \log(100 - F_{15}) - 1.8808)$	BF17	$BF15 \times \max(0, \log(S) + 0.2518)$
		BF18	$\max(0, M - 6.6)$

10.5 Design Charts

The preceding results for both free-face and gentle slope grounds indicate that Eqs. (10.3) and (10.4) can provide reasonable estimates of lateral displacements. Some parametric studies were carried out using the developed MARS model to examine further the influence of F_{15} and $D50_{15}$ on the lateral displacement using the database of only the Niigate case histories ($M = 7.5, R = 21$ km and free-face ground condition with average W of 8.4%, and average T_{15} of 10.3 m). The variations of F_{15} in the range of 2–32%, and $D50_{15} = 0.2$ mm and 0.3 mm were considered, and the MARS predictions are shown in Fig. 10.8. Also plotted in Fig. 10.8 are the measured displacements for $D50_{15}$ values that were close to 0.2 and 0.3 mm. The results show that the MARS model gives logical and consistent trends and indicate that both F_{15} and $D50_{15}$ have an influence on the lateral displacement. The general trend was for D_h value to decrease with increasing F_{15} and with increasing $D50_{15}$.

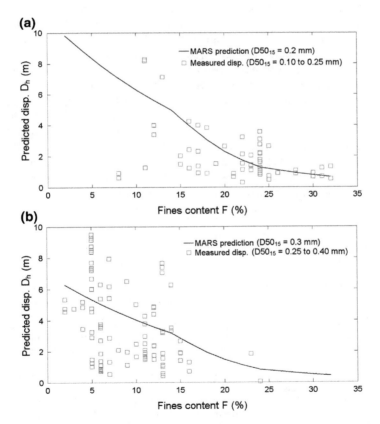

Fig. 10.8 MARS predicted D_h values for Niigate free-face ground: **a** $D50_{15} = 0.2$ mm and **b** $D50_{15} = 0.3$ mm

10.6   Discussions and Summary

MARS model has been presented for estimating the magnitude of liquefaction-induced lateral spread based on database records from case histories provided by Youd et al. (2002), in comparison with the multiple linear regression (MLR) model proposed by Youd et al. (2002). Comparisons indicate that MARS model in predicting liquefaction-induced lateral spread is an improvement over the widely used MLR method. MARS is capable of modeling the nonlinear relationships involving a multitude of variables with interactions among each other without making any specific assumption about the underlying functional relationship between the input variables and the response. The MARS approach is also computationally efficient and is able to provide the relative importance of the input variables. Since MARS explicitly defines the intervals for the input variables, the model enables engineers to have an insight and understanding of where significant changes in the data may occur.

It should be noted that the performance of MARS deteriorates significantly when small or scarce sample sets are used. In addition, as the built MARS model makes predictions based on the knot values and the basis functions, interpolations between the knots of design variables are more accurate and reliable than extrapolations. The proposed MARS model was developed using case history data points with limited ranges of earthquake parameters, soil properties, and geometric attributes. Therefore, it is not recommended that the model be applied for values of input parameters beyond the specified ranges in this study. For this reason, the developed MARS model is incapable of neither assessing the small/little lateral spreading or lateral spreading for claylike sediment, nor correcting the model bias problem, the same disadvantage as the MLR model (Youd et al. 2002), as pointed out by Chu et al. (2006) and Youd et al. (2009). Additional new data sets of little/zero displacement sites are required to further evaluate and update the current MARS model.

References

Al Bawwab WM (2005). Probabilistic assessment of liquefaction-induced lateral ground deformations. Ph.D. Thesis, Department of Civil Engineering, Middle East Technical University

Aydan Ö, Ulusay R, Atak VO (2008) Evaluation of ground deformations induced by the 1999 Kocaeli earthquake (Turkey) at selected sites on shorelines. Environ Geol 54(1):165–182

Bardet JP, Mace N, Tobita T (1999) Liquefaction-induced ground deformation and failure. Report to PEER/PG&E, Task 4A—Phase 1, University of Southern California, Civil Engineering Department

Bardet JP, Tobita T, Mace N, Hu J (2002) Regional modeling of liquefaction-induced ground deformation. Earthq Spectra 18(1):19–46

Bartlett SF, Youd TL (1992a) Empirical analysis of horizontal ground displacement generated by liquefaction-induced lateral spreads. Technical report No. NCEER-92-0021, National Center for Earthquake Engineering Research, State University of New York, Buffalo, NY, 114

Bartlett SF, Youd TL (1992b) Case histories of lateral spreads caused by the 1964 Alaska earthquake. Case Studies of liquefaction and lifeline performance during past earthquakes: Technical report NCEER-92-0002, vol 2, National Center for Earthquake Engineering Research, State University of New York, Buffalo, NY

Bartlett SF, Youd TL (1995) Empirical prediction of liquefaction-induced lateral spread. J Geotech Eng, ASCE 121(4):316–329

Chiru-Danzer M, Juang CH, Christipher RA, Suber J (2001) Estimation of liquefaction-induced horizontal displacements using artificial neural networks. Can Geotech J 38(1):200–207

Chu DB, Stewart JP, Youd TL, Chu BL (2006) Liquefaction-induced lateral spreading in near-fault regions during the 1999 Chi-Chi, Taiwan earthquake. J Geotech Geoenviron Eng, ASCE 132(12):1549–1565

Hamada M, Yasuda S, Isoyama R, Emoto K (1986) Study on liquefaction-induced permanent ground displacement. Report for the Association for the Development of Earthquake Prediction

Kanibir A, Ulusay R, Aydan Ö (2006) Liquefaction-induced ground deformations on a lake shore (Turkey) and empirical equations for their prediction. IAEG2006 Paper number 362, The Geological Society of London

Rauch AF, Martin JR Jr (2000) EPOLLS model for predicting average displacements on lateral spreads. J Geotech Geoenviron Eng, ASCE 126(4):360–371

Shamoto Y, Zhang JM, Tokimatsu K (1998) New charts for predicting large residual post-liquefaction ground deformation. Soil Dyn Earthq Eng 17(7–8):427–438

Wang J, Rahman MS (1999) A neural network model for liquefaction-induced horizontal ground displacement. Soil Dyn Earthq Eng 18(8):555–568

Youd TL, Perkins DM (1987) Mapping of liquefaction severity index. J Geotech Eng, ASCE 113(11):1374–1392

Youd TL, Hanson CM, Bartlett SF (1999) Revised MLR equations for predicting lateral spread displacements. Proceedings of the 7th U.S.-Japan Workshop on Earthquake Resistant Design of Lifeline Facilities and Countermeasures for Soil Liquefaction, MCEER, pp 99–114

Youd TL, Hansen CM, Bartlett SF (2002) Revised multi-linear regression equations for prediction of lateral spread displacement. J Geotech Geoenviron Eng, ASCE 128(12):1007–1017

Youd TL, DeDen DW, Bray JD, Sancio R, Cetin KO, Gerber TM (2009) Zero-displacement lateral spreads, 1999 Kocaeli, Turkey, Earthquake. J Geotech Geoenviron Eng, ASCE 135(1):46–61

Zhang G (2001) Estimation of liquefaction-induced ground deformations by CPT&SPT-based approaches. Ph.D. thesis, University of Alberta, Edmonton, Alberta, Canada

Zhang J, Zhao JX (2005) Empirical models for estimating liquefaction-induced lateral spread displacement. Soil Dyn Earthq Eng 25(6):439–450

Zhang G, Robertson PK, Brachman RWI (2004) Estimating liquefaction-induced lateral displacements using the standard penetration test or cone penetration test. J Geotech Geoenviron Eng, ASCE 130(8):861–871

Chapter 11
MARS Use in Assessment of Soil Liquefaction Based on Capacity Energy Concept

Soil liquefaction is one of the most complicated phenomena to assess in geotechnical earthquake engineering. The procedures that have been developed to determine the liquefaction potential of sandy soil deposits can be categorized into three main groups: stress-based, strain-based, and energy-based procedures. The main advantage of the energy-based approach over the other two methods is the fact that it considers the effects of strain and stress concurrently unlike the stress or strain-based methods. Several liquefaction evaluation procedures have been developed, relating the capacity energy to initial soil parameters such as the relative density, initial effective confining pressure, fine contents, and soil textural properties. Analyses have been carried out on a total of 302 previously published tests using MARS, to assess the capacity energy required to trigger liquefaction in sand and silty sands.

11.1 Background

One of the major causes of damage to the civil engineering structures during earthquakes is due to liquefaction of loose saturated sands and silty sands. Several procedures have been developed to evaluate the liquefaction potential in the field. The available evaluation procedures can be categorized into three main groups: (1) stress-based procedures, (2) strain-based procedures, and (3) energy-based procedures.

The stress-based procedure (Seed and Idriss 1971; Whitman 1971) is the most widely adopted method for liquefaction assessment. The method is mainly empirical and is based on laboratory and field observations. The shear stress level and the number of cycles are the major criteria in this approach. In order to correlate the earthquake actual motion to laboratory harmonic loading conditions, the equivalent stress intensity and the number of cycles have to be defined (Seed and Idriss 1971). Seed et al. (1975) selected the equivalent stress as 65% of the maximum shear stress induced in the earth structure, while Ishihara and Yasuda (1975) proposed 57% rather than 65% for 20 cycles of loading. Some probabilistic frameworks for assessing liquefaction potential of soils based on in situ tests such as the cone penetration test

© Science Press and Springer Nature Singapore Pte Ltd. 2019
W. Zhang, *MARS Applications in Geotechnical Engineering Systems*,
https://doi.org/10.1007/978-981-13-7422-7_11

(CPT) or standard penetration test (SPT) were also carried out (Juang et al. 1999, 2001, 2012; Moss et al. 2006; Boulanger and Idriss 2012). Despite the fact that the stress-based procedure has been continuously revised and extended in subsequent studies and the database of liquefaction case histories expanded, the uncertainty concerning random loading still persists (Green 2001; Baziar and Jafarian 2007).

Dobry et al. (1982) proposed the strain-based procedure as an alternative to the empirical stress-based procedure. This method was derived from the mechanics of two interacting idealized sand grains and then generalized for natural soil deposits (Green 2001; Baziar and Jafarian 2007). It is based on the hypothesis that pore water pressure initiates to develop when the shear strain surpasses a threshold shear strain, which is shown to be approximately 0.01%, irrespective of sand type, relative density, initial effective confining pressure, and sample preparation method. Although this strain-based approach is theoretically reasonable, it is less popular than the stress-based procedure due to the fact that the strain approach only estimates the initiation of pore pressure buildup which is essential for liquefaction to occur, but does not necessarily imply that liquefaction will occur. The main deficiency of this method is the greater difficulty of estimating the cyclic strain compared with the cyclic shear stress (Seed 1980).

Davis and Berrill (1982) introduced an energy-based approach for liquefaction potential assessment in which the energy content of an earthquake is compared with the amount of dissipated energy required for soil liquefaction, known as "capacity energy." The basic elements of both the stress and strain methods are incorporated in the formulation of the energy-based method. The amount of total strain energy at the onset of liquefaction can be obtained from laboratory testing or field records. In a typical cyclic laboratory test, the stress, strain, and pore pressure time histories are recorded. Hysteresis loops can be generated from these stress and strain time histories. Figure 11.1 illustrates a typical hysteresis loop (Green 2001). In other words, this area represents the dissipated energy per unit volume of the soil mass (Ostadan et al. 1996). This is based on the idea that during deformation of cohesionless soils under dynamic loads, part of the energy is dissipated into the soil (Nemat-Nasser and Shokooh 1979). The instantaneous energy and its summation over time intervals are computed until the onset of liquefaction. The summation of the energy at this time is used as the measure of the capacity of the soil sample against initial liquefaction occurrence in terms of the strain energy (capacity energy).

The energy-based approach has the following advantages in comparison with the other existing methods to evaluate the liquefaction potential of soils (Baziar and Jafarian 2007; Baziar et al. 2011):

(1) Energy is associated with the quality of both shear stress and shear strain;
(2) Energy is a scalar quantity which can be associated with the main characterizing earthquake parameters such as source to site distance and magnitude of the earthquake, while it considers the entire spectrum of ground motions as opposed to the stress-based approach, which uses only the peak value of ground acceleration;

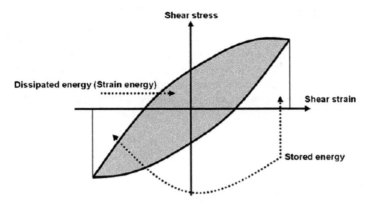

Fig. 11.1 A typical hysteresis shear stress–strain loop (after Green 2001)

(3) It is capable of accounting for the effects of a complicated stress–strain history on pore water pressure buildup.

The energy-based liquefaction evaluation procedures are mainly grouped into approaches developed using earthquake case histories and those developed from laboratory data (Green 2001). Several models were developed relating the soil capacity energy to initial effective mean confining pressure and initial relative density on the basis of a series of laboratory cyclic shear and centrifuge tests (Figueroa et al. 1994; Liang 1995; Dief and Figueroa 2001), and most of these relationships were derived by performing multiple linear regression (MLR) analysis. Although high correlation coefficient R values can be obtained from these models, these relationships were only based on a limited number of tests and failed to take into account the important role of the fines content in the evaluation of the liquefaction behavior. Furthermore, Baziar and Jafarian (2007) demonstrated that such relationships developed based on a limited number of data could not reasonably work in a large data set of various types of sand. Using their compiled database, Baziar and Jafarian (2007) developed a new MLR-based relationship to emphasize the necessity of developing an artificial neural network (ANN)-based model. In addition, Chen et al. (2005) presented a seismic wave energy-based method with back-propagation neural networks to assess the liquefaction potential. Despite the good performance of the ANN-based models, the nature of a black-box framework restricts the practical applications of ANN. Expanding the database collected by Baziar and Jafarian (2007), Baziar et al. (2011) utilized an evolutionary approach based on genetic programming (GP) for the estimation of capacity energy of liquefiable soils. Using the same database as Baziar and Jafarian (2007), Alavi and Gandomi (2012) presented promising variants of GP, namely the linear genetic programming (LGP) and multiexpression programming (MEP) to evaluate the liquefaction resistance of sandy soils. Cabalar et al. (2012) presented an alternative rule-based simulation for the prediction of liquefaction triggering through a novel neuro-fuzzy (NF) approach which possesses the natural language description of fuzzy systems and the learning capability of neural networks.

This chapter adopts the MARS approaches to derive the relationship between the capacity energy dissipated and the soil initial parameters based on a wide-ranging database of laboratory tests. Case histories validating the proposed MARS model were also carried out through comparisons with centrifuge test results. Further, the prediction performance of the derived MARS model compared favorably with other regression and soft computing models.

11.2 The Database

The database used for MARS modeling comprises of 217 cyclic triaxial tests, 61 cyclic torsional shear tests, six cyclic simple shear tests, and 18 centrifuge tests giving a total of 302 tests. A summary of the laboratory tests as well as the parameter statistics is listed in Table 11.1. The specific details of the 302 tests, including the test type, values of parameters, and the failure mode can be found in Baziar and Jafarian (2007).

Of the 302 data sets, 226 (approximately 75% of the total data sets) were randomly selected as the training patterns, while the remaining 76 were used for testing purposes. As the previously proposed regression and soft computing models did not indicate the specific information of the training and testing patterns, for performance comparison, the criterion of data pattern selection used in this study was based on ensuring that the statistical properties including the mean and standard deviations of the training and testing subsets were similar to each other.

Table 11.1 Statistics of parameters used for MARS model development

Parameters	Parameter description	Min.	Max.	Mean	S.D.
Inputs					
σ'_{mean} (kPa)	Initial effective mean confining pressure	27.8	294	94.9	31.3
D_r (%)	Initial relative density after consolidation	−44.5	105.1	49.0	32.6
FC (%)	Percentage of fines content	0	100	20.0	25.9
C_u	Coefficient of uniformity	1.57	5.88	2.42	1.09
D_{50} (mm)	Mean grain size	0.03	0.46	0.23	0.13
Outputs					
$Log(W)$ (J/m^3)	Logarithm of capacity energy	2.48	4.54	3.25	0.45

Min. denotes the minimum value, while Max. represents the maximum value; Mean is the average value, and S.D. denotes the standard deviation

Fig. 11.2 Comparison
between target and MARS
predicted Log(W)

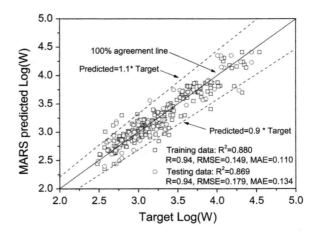

11.3   MARS Modeling Results

The prediction of capacity energy using the MARS model with second-order interaction adopted 14 BFs of linear spline function. The predictions are shown in Fig. 11.2 along with the performance statistics (the coefficient of determination R^2, the coefficient of correlation R, the root mean squared error RMSE, and the mean average error MAE) for the training and testing patterns. It is obvious that the MARS model has been able to learn the complicated relationship between the capacity energy and the soil initial parameters. The relative errors (defined as the ratio of the difference between the MARS predicted and the target logarithmic value of capacity energy Log(W) divided by the target value, in percentage) for the training and testing patterns are plotted in Fig. 11.3. It is obvious that all of the MARS estimations of the data patterns fell within $\pm 20\%$ of the target values, and most of the predictions were within $\pm 10\%$ of the target values.

11.4   Parameter Relative Importance

Table 11.2 displays the ANOVA decomposition of the developed MARS model. The first column lists the ANOVA function number. The second column gives an indication of the importance of the corresponding ANOVA function, by listing the GCV score for a model with all BFs corresponding to that particular ANOVA function removed. The third column provides the standard deviation of the function. It gives an indication of its relative importance to the overall model and can be interpreted in a manner similar to the standardized regression coefficient in a linear model. The fourth column gives the number of BFs comprising the ANOVA function. The last column gives the particular input variables associated with the ANOVA function.

Fig. 11.3 Variation of the relative errors obtained from MARS model

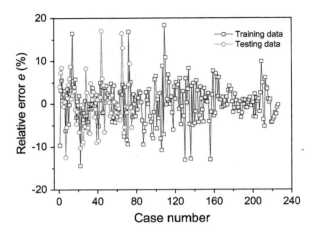

Table 11.2 ANOVA decomposition for MARS model

Function	GCV	STD	#basis	Variable(s)
1	0.056	0.135	2	σ'_{mean}
2	0.039	0.075	1	D_r
3	0.359	0.329	2	FC
4	0.100	0.130	2	C_u
5	0.195	0.249	1	D_{50}
6	0.465	0.410	2	D_r and FC
7	0.036	0.058	1	D_r and D_{50}
8	0.041	0.083	1	FC and D_{50}
9	0.068	0.144	2	C_u and D_{50}

Figure 11.4 gives the plot of the relative importance of the input variables for the MARS model, which is evaluated by the increase in the GCV value caused by removing the considered variables from the developed MARS model. The results indicate that the capacity energy $Log(W)$ is more sensitive to FC compared with the relative density D_r and initial mean effective stress σ'_{mean}, which is consistent with the parametric study of Baziar and Jafarian (2007). According to Baziar and Jafarian (2007) and Alavi and Gandomi (2012), as σ'_{mean} and D_r represent the initial density of the soils, they can be categorized into one group referred to as intergranular contact density. Similarly, C_u and D_{50} are grain size distribution parameters and have been grouped as the grain size characteristics or textural properties. FC is individually considered as a category controlling the potential of pore water pressure buildup. In terms of the three categories, the total relative importance values for intergranular contact density, textural properties, and fines content are 77.2, 103.1, and 100%, respectively. This confirms the finding by Baziar and Jafarian (2007) that the grain size characteristic is the most influential category.

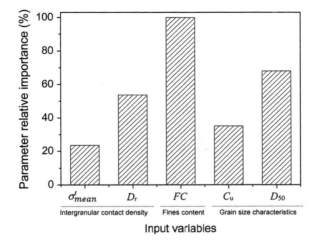

Fig. 11.4 Relative importance of the input variables in the MARS model

Table 11.3 Expressions of BFs for MARS model

BF	Equation	BF	Equation
BF1	$\max(0, D_{50} - 0.12)$	BF8	$\max(0, D_r - 17) \times \max(0, 35 - FC)$
BF2	$BF1 \times \max(0, D_r - 69.2)$	BF9	$BF1 \times \max(0, C_u - 1.68)$
BF3	$\max(0, \sigma'_{mean} - 100.5)$	BF10	$\max(0, FC - 20)$
BF4	$\max(0, 100.5 - \sigma'_{mean})$	BF11	$\max(0, 20 - FC)$
BF5	$\max(0, 17 - D_r)$	BF12	$\max(0, C_u - 2.63)$
BF6	$BF1 \times \max(0, FC - 17)$	BF13	$\max(0, 2.63 - C_u)$
BF7	$\max(0, D_r - 17) \times \max(0, FC - 35)$	BF14	$BF1 \times \max(0, C_u - 1.66)$

11.5 The Developed MARS Model

Table 11.3 lists the BFs and their corresponding equation for the developed MARS model. It is observed from Table 11.3 that interactions have occurred between BFs (6 of the 14 BFs are interaction terms). The presence of interactions suggests that the built MARS model is not simply additive and that interactions play a significant role in building an accurate model for capacity energy predictions. This again indicates that MARS is capable of capturing the nonlinear and complex relationships between energy capacity and a multitude of initial soil parameters with interactions among each other without making any specific assumption about the underlying functional relationship between the input variables and the dependent response. The equation of MARS energy capacity model is given by

$$\begin{aligned}
\mathrm{Log}(W) = {} & 3.28 + 2.11 \times \mathrm{BF1} + 0.057 \times \mathrm{BF2} + 0.0034 \times \mathrm{BF3} - 0.005 \\
& \times \mathrm{BF4} - 0.0074 \times \mathrm{BF5} + 0.11 \times \mathrm{BF6} + 0.00034 \\
& \times \mathrm{BF7} + 0.00038 \times \mathrm{BF8} + 157.14 \times \mathrm{BF9} - 0.018 \\
& \times \mathrm{BF10} - 0.02 \times \mathrm{BF11} - 0.098 \times \mathrm{BF12} - 0.33 \\
& \times \mathrm{BF13} - 156.13 \times \mathrm{BF14}
\end{aligned} \tag{11.1}$$

11.6   Parametric Sensitivity Analysis

To validate the MARS energy capacity model, a parametric analysis was performed, aiming to find the effect of each input variable on the capacity energy. This parametric sensitivity analysis investigates the response of $\mathrm{Log}(W)$ predicted by the MARS model to a set of hypothetical input data generated over the ranges of the minimum and maximum data sets. One input variable was changed each time within its range, while the others were kept at the average values of their entire data sets. As suggested by Alavi et al. (2011), a set of synthetic data for the single varying parameter were generated by increasing the value of this in increments. These values were presented to the MARS prediction model, and $\mathrm{Log}(W)$ was calculated. This procedure was repeated using another variable until the responses of the models were tested for all of the predictor variables (Alavi et al. 2011). Figure 11.5a–e presents the tendency of the $\mathrm{Log}(W)$ predictions to the variations of σ'_{mean}, D_r, FC, C_u, and D_{50}, respectively.

Figure 11.5 indicates that the capacity energy continuously increases due to increasing σ'_{mean}, D_r, and D_{50}, and decreases with increasing C_u. It is easy to understand that an increase in effective confining pressure increases the required energy for liquefaction occurrence and that the capacity energy for the firmer/denser sample is greater. In addition, the increase in capacity energy versus the increasing mean grain size D_{50} indicates that the capacity energy of a coarser soil is higher than that of a finer soil, which was also reported by Liang (1995). The observation that the capacity energy decreases with an increase in C_u may be due to the fact that high C_u values address gap-graded gradations and correspond to high amounts of fines. Figure 11.5c illustrates the results of the parametric study for FC. Compared with the effect of other parameters, the effect of FC on capacity energy is more complicated. There is no clear consensus in the literature about the effect of fines content increment on the trend of capacity energy. While some researchers have shown that an increase in FC decreases the capacity energy, others have reported the opposite effect. Naeini and Baziar (2004) and Xenaki and Athanasopoulos (2003) reported that the liquefaction resistance (capacity energy) decreases when FC increases up to 35 and 44%, and subsequently, the resistance starts increasing. Polito and Martin (2001) showed through a laboratory parametric study that the liquefaction resistance drops as the FC value exceeds the limiting threshold value, and this threshold value is between 25 and 45% for most sands. A similar finding was reported by Alavi and Gandomi (2012). Baziar and Jafarian (2007) showed that the energy-based liquefaction resistance of sand-silt mixtures slightly increases when FC increases up to

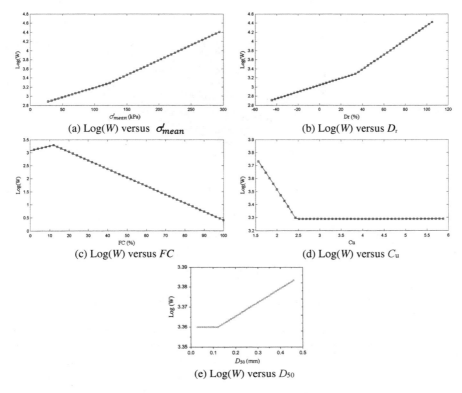

(a) Log(W) versus σ'_{mean}

(b) Log(W) versus D_r

(c) Log(W) versus FC

(d) Log(W) versus C_u

(e) Log(W) versus D_{50}

Fig. 11.5 Parametric analysis of the capacity energy by MARS model

10% and then continuously decreases while its decrement rate declines for greater FC. Baziar et al. (2011) proposed that capacity energy initially increases versus the increment of FC from zero to about 10–15%, and then, the trend is reversed for further increases in FC. This MARS capacity energy model is in agreement with the findings from Baziar et al. (2011) that indicate that the peak capacity energy is obtained when the FC value is between 10 and 15%.

The influence of input parameters on the capacity energy by the MARS model can be illustrated graphically. The variation of capacity energy is plotted against σ'_{mean} and D_r in Fig. 11.6. A similar plot was presented in Cabalar et al. (2012) through the ANFIS model. It is obvious that the MARS model is more physically reasonable since larger σ'_{mean} and D_r values result in greater capacity energy. However, Cabalar et al. (2012) reported an opposite observation.

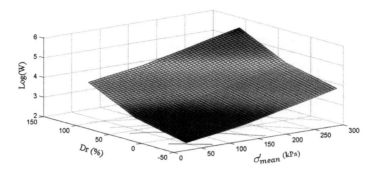

Fig. 11.6 Nonlinear mapping surface relating Log(W) to σ'_{mean} and D_r

11.7 Validation from Case Histories

To further validate the proposed MARS energy capacity model, the results of 22 centrifuge tests on different sand samples (Dief 2000) as listed in Table 11.4 were analyzed. Also listed in Table 11.4 are the MARS predicted Log(W) values and the relative error e (defined as the ratio of the difference between the MARS predicted and the target Log(W) divided by the target value, in percentage). Figure 11.7 presents the histogram plots of the relative errors. It is obvious that most of the MARS estimations of the data patterns fell within $\pm10\%$ of the target values, indicating that the proposed MARS energy capacity model is able to predict reasonably well the capacity energy required to trigger liquefaction in sands.

Table 11.4 Summary of centrifuge data sets for validation of the proposed MARS energy capacity model (after Dief 2000)

Test No.	σ'_{mean} (kPa)	D_r (%)	FC (%)	C_u	D_{50} (mm)	Target Log(W)	MARS predicted Log(W)	e (%)
Nevada sand								
1	28.7	60.7	0	2.27	0.15	2.77	2.95	6.48
2	33.9	58.5	0	2.27	0.15	2.78	2.95	6.16
3	28.4	64.7	0	2.27	0.15	2.89	3.00	3.69
4	34.4	66.5	0	2.27	0.15	2.97	3.06	2.92
5	29.8	72.0	0	2.27	0.15	2.97	3.11	4.78
6	34.7	72.1	0	2.27	0.15	3.04	3.14	3.38
7	28.8	76.3	0	2.27	0.15	3.15	3.17	0.68

(continued)

Table 11.4   (continued)

Test No.	σ'_{mean} (kPa)	D_r (%)	FC (%)	C_u	D_{50} (mm)	Target Log(W)	MARS predicted Log(W)	e (%)
Reid Bedford sand								
8	28.8	51.0	0	1.67	0.26	2.74	2.72	−0.88
9	33.6	51.0	0	1.67	0.26	2.85	2.74	−3.83
10	34.1	60.2	0	1.67	0.26	2.95	2.87	−2.78
11	29.1	71.8	0	1.67	0.26	3.05	3.01	−1.18
12	29.3	78.7	0	1.67	0.26	3.06	3.16	3.14
13	34.6	80.4	0	1.67	0.26	3.23	3.23	−0.01
LSFD sand								
14	14.1	55.0	28	5.88	0.13	2.59	3.23	24.7
15	31.3	62.8	28	5.88	0.13	2.66	2.61	−1.93
16	16.1	67.2	28	5.88	0.13	2.60	2.61	0.32
17	31.7	67.2	28	5.88	0.13	2.68	2.63	−1.93
18	15.2	72.0	28	5.88	0.13	2.62	2.63	0.31
19	31.5	72.0	28	5.88	0.13	2.70	2.64	−2.31
20	15.4	86.2	28	5.88	0.13	2.63	2.64	0.38
21	32.3	88.5	28	5.88	0.13	2.71	2.70	−0.38
22	15.7	93.0	28	5.88	0.13	2.64	2.70	2.28

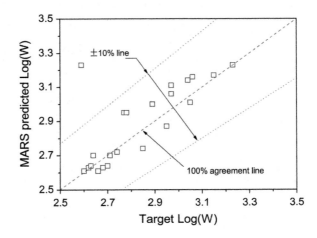

Fig. 11.7   MARS predicted Log(W) versus target Log(W) from centrifuge tests

11.8 Performance Comparisons

The overall performance statistics of the models obtained by MARS, linear genetic programming LGP, multiexpression programming MEP, standard genetic programming GP, adaptive neuro-fuzzy inference system ANFIS, and the conventional multiple linear regression MLR-based equations are summarized in Table 11.5. It is clearly observed that the MARS model is able to predict the capacity energy reasonably well compared with other regression and soft computing algorithms including artificial neural network and genetic algorithm. It is not possible to assess which method is more accurate or reasonable since the training and testing patterns used for model development for each method are different. The simple MLR model proposed by Baziar and Jafarian (2007) is not as accurate as other models since it only comprised of the linear terms of inputs and thus failed to capture the nonlinear relationships involving a multitude of variables with interaction among each other. The more advanced algorithms such as GP, ANN, AIFIS, LGP, and MEP are capable of extracting the functional underlying relationships based on the training cases. However, there is a lack of interpretation as to physical meaning of the generalized weight and bias values. In addition, these techniques generally require additional data preprocessing such as "scaling" or "normalization" as well as post-processing or "denormalization." An obvious advantage of the MARS model lies in the fact that it is much easier to interpret, as shown in Eq. (11.1) and Table 11.3. Furthermore, it does not need any normalization process. Nevertheless, all these models listed in Table 11.5 give comparably accurate estimations and can be used for cross-validating each other.

Table 11.5 Summary of performance measures of the energy-based models for liquefaction assessment

Model	No. of data sets	Performances				Reference
		R^2	R	RMSE	MAE	
MLR	284	0.65	–	0.262	0.213	Baziar and Jafarian (2007)
ANN	284	0.90	–	0.138	0.104	Baziar and Jafarian (2007)
GP	399	0.88	–	0.140	0.109	Baziar et al. (2011)
LGP	301	–	0.87	0.224	0.178	Alavi and Gandomi (2012)
MEP	301	–	0.86	0.233	0.187	Alavi and Gandomi (2012)
GP	301	–	0.81	0.274	0.219	Alavi and Gandomi (2012)
ANFIS	302	0.87	–	0.181	–	Cabalar et al. (2012)
MARS	302	0.88	0.94	0.182	0.155	This chapter

"–" indicates that this performance statistics is not provided in the reference

11.9 Summary

The proposed MARS model estimations of the capacity energy are comparable with other models developed using soft computing algorithms. The developed MARS model gives predictions that are just as accurate as other soft computing techniques. Additional centrifuge case histories demonstrating the reliability of the proposed MARS capacity energy model are also given.

It should be noted that since the built MARS model makes predictions based on the knot values and the basis functions; thus, interpolations between the knots of design input variables are more accurate and reliable than extrapolations. Consequently, it is not recommended that the model be applied for values of input parameters beyond the specific ranges in this chapter.

References

Alavi AH, Gandomi AH (2012) Energy-based numerical models for assessment of soil liquefaction. Geosci Front 3(4):541–555

Alavi AH, Ameri M, Gandomi AH, Mirzahosseini MR (2011) Formulation of flow number of asphalt mixes using a hybrid computational method. Constr Build Mater 25:1338–1355

Baziar MH, Jafarian Y (2007) Assessment of liquefaction triggering using strain energy concept and ANN model capacity energy. Soil Dyn Earthq Eng 27:1056–1072

Baziar MH, Jafarian Y, Shahnazari H, Movahed V, Tutunchian MA (2011) Prediction of strain energy-based liquefaction resistance of sand-silt mixtures: an evolutionary approach. Comput Geosci 37(11):1883–1893

Boulanger RW, Idriss IM (2012) Probabilistic standard penetration test-based liquefaction-triggering procedure. J Geotech Geoenviron 138:1185–1195

Cabalar AF, Cevik A, Gokceoglu C (2012) Some applications of Adaptive Neuro-Fuzzy Inference System (ANFIS) in geotechnical engineering. Comput Geotech 40:14–33

Chen YR, Hsieh SC, Chen JW, Shih CC (2005) Energy-based probabilistic evaluation of soil liquefaction. Soil Dyn Earthq Eng 25(1):55–68

Davis RO, Berrill JB (1982) Energy dissipation and seismic liquefaction in sands. Earthq Eng Struct D 10:59–68

Dief HM (2000) Evaluating the liquefaction potential of soils by the energy method in the centrifuge. Ph.D. Dissertation, Department of Civil Engineering, Case Western Reserve University, Cleveland, OH

Dief HM, Figueroa JL (2001) Liquefaction assessment by the energy method through centrifuge modeling. In: Zeng XW (ed) Proceedings of the NSF International Workshop on Earthquake Simulation in Geotechnical Engineering, CWRU, Cleveland, OH

Dobry R, Ladd RS, Yokel FY, Chung RM, Powell D (1982) Prediction of pore water pressure build-up and liquefaction of sands during earthquakes by the cyclic strain method. National Bureau of Standards, US Department of Commerce, US Governmental Printing Office, Building Science Series, Washington, DC

Figueroa JL, Saada AS, Liang L, Dahisaria MN (1994) Evaluation of soil liquefaction by energy principles. J Geotech Geoenviron 20(9):1554–1569

Green RA (2001) Energy-based evaluation and remediation of liquefiable soils. Ph.D. dissertation, Virginia Polytechnic Institute and State University, Blacksburg, VA

Ishihara K, Yasuda S (1975) Sand liquefaction in hollow cylinder torsion under irregular excitation. Soils Found 15(1):45–59

Juang CH, Rosowsky DV, Tang WH (1999) Reliability-based method for assessing liquefaction potential of soils. J Geotech Geoenviron 125:684–689

Juang CH, Chen CJ, Jiang T (2001) Probabilistic framework for liquefaction potential by shear wave velocity 127:670–678

Juang CH, Ching J, Luo Z, Ku CS (2012) New models for probability of liquefaction using standard penetration tests based on an updated database of case histories 133–134:85–93

Liang L (1995) Development of an energy method for evaluating the liquefaction potential of a soil deposit. Ph.D. thesis, Department of Civil Engineering, Case Western Reserve University, Cleveland, OH

Moss RES, Seed RB, Kayen RE, Stewart JP, Der Kiureghian AK. Cetin KO (2006) CPT-based probabilistic and deterministic assessment of in situ seismic soil liquefaction potential. J Geotech Geoenviron Eng ASCE 132(8):1032–1051

Naeini SA, Baziar MH (2004) Effect of fines content on steady-state strength of mixed and layered samples of a sand. Soil Dyn Earthq Eng 24:181–187

Nemat-Nasser S, Shokooh A (1979) A unified approach to densification and liquefaction of cohesionless sand in cyclic shearing. Can Geotech J 16(4):659–678

Ostadan F, Deng N, Arango I (1996) Energy-based method for liquefaction potential evaluation, phase I. feasibility study. U.S. Department of Commerce, Technology Administration, National Institute of Standards and Technology, Building and Fire Research Laboratory

Polito CP, Martin JR (2001) Effects of non-plastic fines on the liquefaction resistance of sands. J Geotech Geoenviron 127(5):408–415

Seed HB (1980) Closure to soil liquefaction and cyclic mobility evaluation for level ground during earthquakes. J Geotech Eng-ASCE 106(GT6):724

Seed HB, Idriss IM (1971) Simplified procedure for evaluating soil liquefaction potential. J Soil Mech Found Div 97(9):1249–1273

Seed HB, Idriss IM, Makdisi F, Banerjee N (1975) Representation of irregular stress time histories by equivalent uniform stress series in liquefaction analyses. Report No. UCB/EERC-75/29, Earthquake Engineering Research Centre, U.C. Berkeley

Whitman RV (1971) Resistance of soil to liquefaction and settlement. Soils Found 11(4):59–68

Xenaki VC, Athanasopoulos GA (2003) Liquefaction resistance of sand-silt mixtures: an experimental investigation of the effect of fines. Soil Dyn Earthq Eng 23:183–194

Chapter 12
MARS_LR Use in Assessment of Soil Liquefaction

Simplified techniques based on in situ testing methods are commonly used to assess seismic liquefaction potential. Many of these simplified methods were developed by analyzing liquefaction case histories from which the liquefaction boundary (limit state) separating two categories (the occurrence or non-occurrence of liquefaction) is determined. As the liquefaction classification problem is highly nonlinear in nature, it is difficult to develop a comprehensive model using conventional modeling techniques that take into consideration all the independent variables, such as the seismic and soil properties. In this chapter, a modification MARS approach based on logistic regression (LR) MARS_LR is used to evaluate seismic liquefaction potential based on actual field records. Three different MARS_LR models were used to analyze three different field liquefaction databases, and the results are compared with BPNN.

12.1 Background

Simplified techniques based on an in situ testing measurement index are commonly used to assess seismic liquefaction potential. Most of these simplified charts or equations rely on the analysis of liquefaction case histories. Using empirical, simple regression, or statistical methods, a boundary (liquefaction curve) or classification technique is used to separate the occurrence or non-occurrence of liquefaction.

Techniques using the standard penetration test (SPT) have been developed for evaluating soil liquefaction potential (Seed and Idriss 1971; Seed et al. 1985; Law et al. 1990; Cetin et al. 2004). Similarly, methods based on the use of the cone penetration test (CPT) have been developed (Stark and Olson 1995; Robertson and Wride 1998; Juang et al. 2003; Moss et al. 2006). Other in situ test methods to evaluate liquefaction potential include the use of the dilatometer (Marchetti 1982) and the shear wave velocity test (Andrus and Stokoe 2000). Statistical methods were commonly adopted to assign probabilities of liquefaction through various statistical classifications and regression analyses (Liao et al. 1988; Juang et al. 1999; Lai et al. 2004; Tosun et al. 2011).

© Science Press and Springer Nature Singapore Pte Ltd. 2019   187
W. Zhang, *MARS Applications in Geotechnical Engineering Systems*,
https://doi.org/10.1007/978-981-13-7422-7_12

Finding the liquefaction boundary separating two categories (the occurrence or non-occurrence of liquefaction) for multivariate variables can be considered as a pattern classification problem. In mathematical terms, an input vector of variables is used to determine a category (classification) by being shown data of known classifications. Some common pattern recognition tools include discriminant analysis (DA) (Friedman 1989), classification and regression tree (CART) (Breiman 1984), neural networks (Specht 1990; Zhang 2000), and support vector machine (SVM) (Vapnik et al. 1997). This study utilizes a modified multivariate adaptive regression spline (MARS) method (Friedman 1991), in which logistic regression (LR) is applied to separate data into various categories.

This chapter adopts the MARS_LR approach to analyze three different databases of field liquefaction CPT case records. These three database case records are from Goh (2002), Juang et al. (2003) Chern et al. (2008), respectively. Each database is used to train and test the reliability of the MARS_LR model to correctly classify the occurrence or non-occurrence of liquefaction, in comparison with the results from the neural network approaches, including the probabilistic neural network (PNN) model proposed by Goh (2002), a three-layered feed-forward network adopted by Juang et al. (2003) and a fuzzy-neural system developed by Chern et al. (2008). For the neural networks, the training data are used to optimize the connection weights to reduce the errors between the actual and target outputs through minimization of the defined error function (e.g., sum squared error) using the gradient descent approach. Validation of the neural network performance is performed by "testing" with a separate set of data that were never used in training process, to assess the generalization capacity of the trained model to produce the correct input–output mapping even the input is different from the datasets those had been used to train the network.

12.2   Modeling Accuracy

Two simple and common methods of evaluating the performance of a pattern classification model are to determine the error rate (the percentage of misclassified cases, termed as ER) or the success rate (the percentage of correctly classified cases, termed as SR). In assessing the performance of various seismic liquefaction potential models, most researchers have either adopted the success rate or error rate as the criterion.

However, the use of either ER or SR does not take into consideration the misclassification costs (classifying liquefied as non-liquefied and non-liquefied as liquefied) which may not be equal or could be subject to change. When the misclassification costs are not equal, then a confusion matrix is commonly used to quantify the costs and minimize the expected loss. A confusion matrix is a table used to evaluate the performance of a classifier. It is a matrix of the observed versus the predicted classes, with the observed classes in columns and the predicted classes in rows as shown in Table 12.1.

Table 12.1 Confusion matrix

		True class	
		Liquefied	Non-liquefied
Predicted class	Liquefied	a	b
	Non-liquefied	c	d

Table 12.1 represents a confusion matrix, where each cell contains a count of seismic liquefaction cases belonging to each particular class. There are four classes in total with each cell labeled by a, b, c, and d. The diagonal elements a and d include the frequencies of correctly classified instances, and the non-diagonal elements b and c include the frequencies of misclassification. The modeling inaccuracy is easily calculated as $\frac{b+c}{a+b+c+d}$, while the modeling accuracy is expressed as $\frac{a+d}{a+b+c+d}$. Other measures of interest are the proportion of liquefied classified as non-liquefied (termed as Type I error), $\frac{c}{a+c}$, and the proportion of non-liquefied classified as liquefied (termed as Type II error), $\frac{b}{b+d}$. In general, the misclassification costs of liquefaction potential associated with Type I error are higher than those associated with Type II error. It is worse to assess a case as non-liquefied when it is actually liquefied than it is to assess a case as liquefied when it is in fact non-liquefied.

12.3 The Databases

12.3.1 Database 1

The database used by Juang et al. (2003) consists of 226 cases, 133 liquefied cases, and 93 non-liquefied. These cases are derived from CPT measurements at over 52 sites and field observations of 6 different earthquakes. The depths h at which the cases are reported range from 1.4 to 14.1 m. For the details of these cases and the neural network approach, the reader is referred to Juang et al. (2003).

The neural network model adopted by Juang et al. (2003) utilizes four input neurons representing normalized core penetration resistance q_{c1N}, the soil type index I_c, the effective stress σ'_v, and the cyclic stress ratio $CSR_{7.5}$. Among the four inputs, σ'_v is the only variable derived directly from CPT measurements. The q_{c1N}, I_c, and $CSR_{7.5}$ are intermediate parameters, determined through the following empirical equations (Juang et al. 2003; Liao et al. 1988; Youd et al. 2001; Robertson and Wride 1998; Juang et al. 2003)

$$CSR_{7.5} = 0.65(\frac{\sigma_v}{\sigma'_v})(\frac{a_{max}}{g})(r_d)/MSF \qquad (12.1)$$

Table 12.2 Summary of CPT databases of field liquefaction cases and neural network modeling results

Database description and neural network modeling results	Database number		
	1	2	3
Observations	226 cases 133 liquefied 93 non-liquefied	170 cases 104 liquefied 66 non-liquefied	466 cases 250 liquefied 216 non-liquefied
Reference	Juang et al. (2003)	Goh (2002)	Chern et al. (2008)
Neural Network structure	Three-layer feed-forward with three hidden neurons	Four-layer PNN with Bayesian classifier	Fuzzy-neural system with four clusters and six hidden neurons
Data patterns for modeling	151 cases for tr. 75 cases for te.	114 cases for tr. 56 cases for te.	350 cases for tr. 116 cases for te.
Modeling results	SR for tr.: 98% SR for te.: 91% Overall SR: 96%	SR for tr.: 100% SR for te.: 100% Overall SR: 100%	SR for tr.: 98% SR for te.: 95.7% Overall SR: 97.4%
Confusion matrix	96.2% | 4.3%	100% | 0%	96.8% | 1.9%
	3.8% | 95.7%	0% | 100%	3.2% | 98.1%

$$r_d = \begin{cases} 1.0 - 0.00765z & z \le 9.15 \text{ m} \\ 1.174 - 0.0267z & 9.15 \text{ m} < z \le 23 \text{ m} \end{cases} \tag{12.2}$$

$$\text{MSF} = 10^{2.24}/M_w^{2.56} = (M_w/7.5)^{-2.56} \tag{12.3}$$

$$q_{c1N} = \frac{q_c/100}{(\sigma_v'/100)^{0.5}} \tag{12.4}$$

$$I_c = [(3.47 - \log_{10} q_{c1N})^2 + (\log_{10} F + 1.22)^2]^{0.5} \tag{12.5}$$

$$F = f_s/(q_c - \sigma_v) \tag{12.6}$$

where f_s = sleeve friction; σ_v = total vertical stress; a_{max} = the peak acceleration at the ground surface; g = acceleration of gravity; r_d = shear stress reduction factor; MSF = magnitude scaling factor; z = depth in meters; M_w = moment magnitude; q_c = measured cone tip resistance; F = normalized friction ratio. The trained neural network structure and modeling results for the training (tr.) and testing (te.) data are summarized in row 3 of Table 12.2.

12.3.2 Database 2

The case records used by Goh (2002) represent 104 sites that liquefied and 66 sites that did not liquefy. PNN approach based on the Bayesian classifier method is used with four layers: the input layer, the pattern layer, the summation layer, and the output layer. The inputs consisted of six neurons representing the earthquake magnitude M, a_{max}, σ_v, σ'_v, q_c, and the mean grain size D_{50}. The trained neural network structure and modeling results are summarized in row 4 of Table 12.2. For the details of these cases and the PNN approach, the reader is referred to Goh (2002).

12.3.3 Database 3

Database 3 compiled by Chern et al. (2008) includes 466 CPT-based field lique-faction records from more than 11 major earthquakes between 1964 and 1999. The records comprised 250 liquefied cases and 216 non-liquefied cases. Chern et al. (2008) developed a fuzzy-neural network to evaluate the liquefaction potential using five parameters: M, σ_v, σ'_v, q_c, and a_{max}. The best-trained neural network model and the modeling results are summarized in row 5 of Table 12.2. In addition, para-metric sensitivity analyses indicated that a_{max} and q_c were the two most important parameters influencing liquefaction assessment.

12.4 The Developed MARS_LR Models and Modeling Results

Three different MARS models were used to analyze the same databases, and the results are compared with the neural network results. Table 12.3 shows the results of MARS models. It summarizes the database used, the input variables, the data sets for training and testing, the model settings, the execution time (PC with 3.0 GHz Intel Core2Quad Q9650 processor, 4 GB RAM), the plot of predictions, the success rates, the confusion matrix, and the basis function together with the performance functions for each MARS model. Figures 12.1, 12.2, and 12.3 illustrate the training and testing results for MARS models, respectively. Tables 12.4, 12.6, and 12.7 list the corresponding basis functions and MARS expressions for these models. Table 12.5 shows the ANOVA decomposition of the various MARS models.

Table 12.3 MARS models and modeling results

Model	I	II	III
Database No.	1	2	3
Input variables	$M\ h\ q_c\ R_f\ \sigma'_v\ \sigma_v\ a_{max}$	$M\ \sigma_v\ \sigma'_v\ q_c\ a_{max}\ D_{50}$	$M\ h\ \sigma_v\ \sigma'_v\ q_c\ a_{max}$
Data sets for tr. and te.	170 for training 56 for testing	114 for training 56 for testing	350 for training 116 for testing
Model settings	13 BFs, second-order interaction, linear spline	6 BFs, second-order interaction, linear spline	12 BFs, second-order interaction, linear spline
Exec. time (s)	0.95	0.17	2.34
Results plot	Fig. 12.1	Fig. 12.2	Fig 12.3
SR	tr.: 94.1% te.: 89.3% Overall: 92.9%	tr.: 90.4% te.: 91.1% Overall: 90.6%	tr.: 93.4% te.: 87.9% Overall: 92.1%
Confusion matrix	130 (97.7%)	14 (15.1%)	99 (95.2%)	11 (16.7%)	227 (90.8%)	14 (6.5%)
	3 (2.3%)	79 (84.9%)	5 (4.8%)	55 (83.3%)	23 (9.2%)	202 (93.5%)
BFs and model expression	Table 12.4	Table 12.6	Table 12.7

12.4.1 MARS_LR Model 1

Model I is the MARS model used to analyze database I which consisted of 170 training and 56 testing records. The input variables were M, h, q_c, R_f, σ'_v, σ_v, and a_{max}. The training and testing results are shown in Fig. 12.1. Model I has an overall success rate of 92.5%. The model accuracy in predicting liquefied cases is very high (97.7%). Type I error is very low (2.3%). Table 12.4 shows the basis function expressions and Model I expression. The derived $f(x)$ can be used to determine the liquefaction potential. ANOVA decomposition of Model I in row 2 of Table 12.5 indicates that q_c and a_{max} are the two most significant parameters. The ANOVA decomposition also indicates that interaction between q_c and a_{max} is significant.

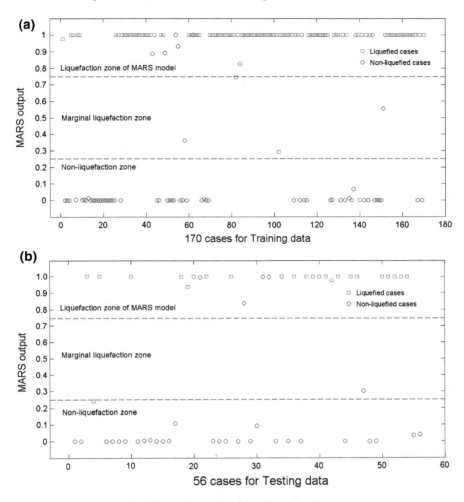

Fig. 12.1 Modeling results of Model I: **a** training data; **b** testing data

12.4.2 MARS_LR Model 2

Model II is the MARS model used to analyze database 2 which consisted of 114 training and 56 testing records. The input variables were M, σ_v, σ'_v, q_c, a_{max}, and D_{50}. The training and testing results are shown in Fig. 12.2. Model II has an overall success rate of 90.6%. The model accuracy in predicting liquefied cases is relatively high (95.2%). Type I error is 4.8%. Table 12.6 shows the basis function expressions and Model II expression. ANOVA decomposition of Model II in row 3 of Table 12.5 indicates that q_c and M are the two most significant parameters. The interaction between M and σ_v is also of significance in assessing liquefaction potential.

Table 12.4 BFs and MARS expression for Model I

Basis functions	Expression
BF1	$\max(0, 0.21 - a_{\max})$
BF2	$\max(0, q_c - 3.1)$
BF3	$\max(0, R_f - 2)$
BF4	$\text{BF2} \times \max(0, 5.8 - h)$
BF5	$\text{BF2} \times \max(0, 0.6 - R_f)$
BF6	$\text{BF2} \times \max(0, a_{\max} - 0.15)$
BF7	$\text{BF2} \times \max(0, 0.15 - a_{\max})$
BF8	$\text{BF3} \times \max(0, a_{\max} - 0.19)$
BF9	$\text{BF3} \times \max(0, 0.19 - a_{\max})$
BF10	$\text{BF1} \times \max(0, M - 6.6)$
BF11	$\max(0, R_f - 2.8)$
BF12	$\max(0, h - 4.2)$
BF13	$\max(0, 4.2 - h)$

$$
\begin{aligned}
y = {} & 26.51 - 372.26 \times \text{BF1} - 5.32 \times \text{BF2} - 40.3 \times \text{BF3} - 0.87 \times \text{BF4} \\
& + 6.89 \times \text{BF5} + 11.95 \times \text{BF6} + 107.85 \times \text{BF7} + 47.23 \times \text{BF8} \\
& + 261.62 \times \text{BF9} + 407.94 \times \text{BF10} + 27.69 \times \text{BF11} \\
& - 1.82 \times \text{BF12} - 5.13 \times \text{BF13}
\end{aligned}
$$

$$f(x) = \frac{1}{1 + e^{-y}}$$

Table 12.5 Parameters derived from ANOVA decomposition for each model

Model	The two most important single variables	The most important interaction terms
I	q_c and $a_{\max}$	$(q_c, a_{\max})$
II	q_c and M	(M, σ_v)
III	q_c and $a_{\max}$	$(q_c, a_{\max})$

Table 12.6 BFs and MARS expression for Model II

Basis functions	Expression
BF1	$\max(0, 13.85 - q_c)$
BF2	$\text{BF1} \times \max(0, 0.16 - a_{\max})$
BF3	$\max(0, M - 6.4)$
BF4	$\max(0, 6.4 - M)$
BF5	$\text{BF3} \times \max(0, 215.7 - \sigma_v)$
BF6	$\text{BF1} \times \max(0, \sigma_v - 153)$

$$f(x) = \frac{1}{1 + e^{-(-22.65 + 3.44 \times \text{BF1} - 72.35 \times \text{BF2} + 13.46 \times \text{BF3} - 32.19 \times \text{BF4} - 0.07 \times \text{BF5} - 0.04 \times \text{BF6})}}$$

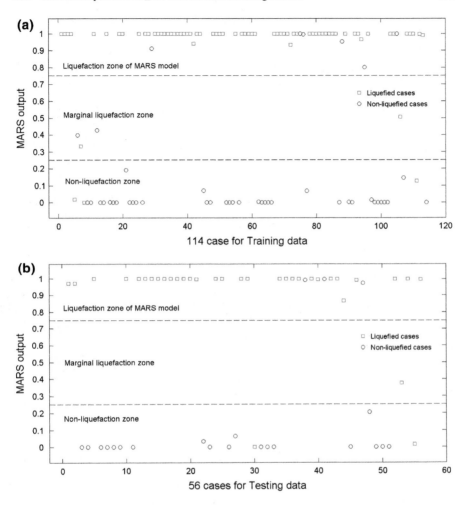

Fig. 12.2   Modeling results of Model II: **a** training data; **b** testing data

12.4.3   MARS_LR Model 3

Model III is the MARS model used to analyze database 3 which consisted of 350 training and 116 testing patterns. The input variables include M, h, σ_v, σ'_v, q_c, and a_{max}. The training and testing results are shown in Fig. 12.3. Model III has an overall success rate of 92.1%. The model accuracy in predicting liquefied cases is 90.8% and in predicting non-liquefied cases is 93.5%. Type I error is 9.2%. Table 12.7 shows the basis function expressions and Model III expression. ANOVA decomposition of Model III in row 4 of Table 12.5 indicates that q_c and a_{max} are the two most

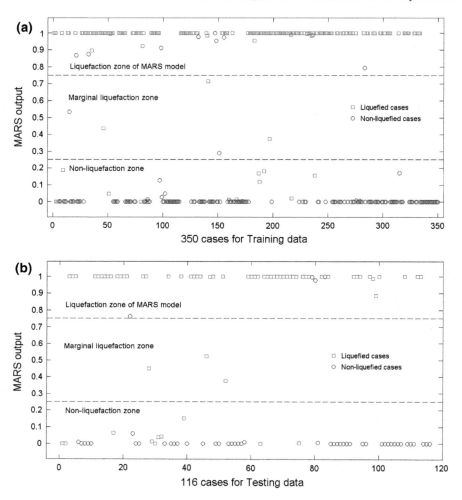

Fig. 12.3 Modeling results of Model III: **a** training data; **b** testing data

significant parameters, which are consistent with the conclusions of Chern et al (2008). The interaction between q_c and a_{max} is also of significance.

12.5   Summary

MARS_LR is the modified MARS approach which is proposed by the author to separate the two categories (the occurrence or non-occurrence of liquefaction). Based on three different databases from the literatures, three different MARS_LR mod-

Table 12.7 BFs and MARS expression for Model III

Basis functions	Expression
BF1	$\max(0, 10.61 - q_c)$
BF2	$\max(0, 0.22 - a_{max})$
BF3	$BF1 \times \max(0, \sigma_v - 136.8)$
BF4	$\max(0, M - 6)$
BF5	$\max(0, 6 - M)$
BF6	$\max(0, a_{max} - 0.22) \times \max(0, 5.5 - q_c)$
BF7	$BF1 \times \max(0, a_{max} - 0.25)$
BF8	$BF1 \times \max(0, 0.25 - a_{max})$
BF9	$BF2 \times \max(0, \sigma'_v - 59.8)$
BF10	$BF2 \times \max(0, 59.8 - \sigma'_v)$
BF11	$\max(0, q_c - 5.8)$
BF12	$\max(0, 5.8 - h)$

$$y = -17.71 + 5.48 \times BF1 - 187.87 \times BF2 - 0.04 \times BF3 + 5.65 \times BF4$$
$$- 202.65 \times BF5 - 27.72 \times BF6 + 9.23 \times BF7 - 22.17 \times BF8$$
$$+ 3.41 \times BF9 + 12.28 \times BF10 - 1.46 \times BF11 - 2.89 \times BF12$$

$$f(x) = \frac{1}{1 + e^{-y}}$$

els were used to analyze the different field liquefaction databases and the results are compared with the neural network approaches. Comparisons indicate that the presented MARS_LR approach in liquefaction assessment use is promising.

References

Andrus RD, Stokoe KH (2000) Liquefaction resistance of soils from shear-wave velocity. J Geotech Geoenviron Eng 126(11):1015–1025

Breiman L, Friedman JH, Olshen RA, Stone CJ (1984) Classification and regression trees. Wadsworth & Brooks, Monterey, CA

Cetin KO, Seed RB, Der Kiureghian AK, Tokimatsu K, Harder LF Jr, Kayen RE, Moss RES (2004) Standard penetration test-based probabilistic and deterministic assessment of seismic soil liquefaction potential. J Geotech Geoenviron Eng ASCE 130(12):1314–1340

Chern SG, Lee CY, Wang CC (2008) CPT-based liquefaction assessment by using fuzzy-neural network. J Mar Sci Technol 16(2):139–148

Friedman JH (1989) Regularized discriminant analysis. J Am Stat Assoc 84(405):165–175

Friedman JH (1991) Multivariate adaptive regression splines. Ann Stat 19:1–141

Goh ATC (2002) Probabilistic neural network for evaluating seismic liquefaction potential. Can Geotech J 39:219–232

Juang CH, Chen CJ (1999) CPT-based liquefaction evaluation using artificial neural networks. Comput Aided Civ Infrastruct Eng 14(3):221–229

Juang CH, Yuan H, Lee DH, Lin PS (2003) Simplified cone penetration test-based method for evaluating liquefaction resistance of soils. J Geotech Geoenviron Eng ASCE 129(1):66–80

Lai SY, Hsu SC, Hsieh MJ (2004) Discriminant model for evaluating soil liquefaction potential using cone penetration test data. J Geotech Geoenviron Eng ASCE 130(12):1271–1282

Law KT, Cao YL, He GN (1990) An energy approach for assessing seismic liquefaction potential. Can Geotech J 27:320–329

Liao SC, Veneziano D, Whitman RV (1988) Regression models for evaluating liquefaction probability. J Geotech Eng ASCE 114(4):389–411

Marchetti S (1982) Detection of liquefiable sand layers by means of quasi-static penetration tests. In: Proceedings of the 2nd European symposium on penetration testing, vol 2, Amsterdam, pp 458–482

Moss RES, Seed RB, Kayen RE, Stewart JP, Der Kiureghian AK, Cetin KO (2006) CPT-based probabilistic and deterministic assessment of in situ seismic soil liquefaction potential. J Geotech Geoenviron Eng ASCE 132(8):1032–1051

Robertson PK, Wride CE (1998) Evaluating cyclic liquefaction potential using the cone penetration test. Can Geotech J 35(3):442–459

Seed HB, Idriss IM (1971) Simplified procedure for evaluating soil liquefaction potential. J Soil Mech Found Div 97(9):1249–1273

Seed HB, Tokimatsu K, Harder LF, Chung R (1985) Influence of SPT procedures in soil liquefaction resistance evaluations. J Geotech Eng ASCE 111(12):861–878

Specht D (1990) Porbabilistic neural networks. Neural Netw 3:109–118

Stark TD, Olson SM (1995) Liquefaction resistance using CPT and field case histories. J Geotech Eng ASCE 121(12):856–869

Tosun H, Seyrek E, Orhan A, Savas H, Turkoz M (2011) Soil liquefaction potential in Eskisehir, NW Turkey. Nat Hazards Earth Syst Sci 11:1071–1082

Vapnik V, Golowich S, Smola A (1997) Support vector method for function approximation, regression estimation, and signal processing. In: Mozer M, Jordan M, Petsche T (eds) Advances in neural information processing systems, vol 9. MIT Press, Cambridge, MA, pp 281–287

Youd TL et al (2001) Liquefaction resistance of soils: Summary report from the 1996 NCEER and 1998 NCEER/NSF workshops on evaluation of liquefaction resistance of soils. J Geotech Geoenviron Eng ASCE 127(10):817–833

Zhang GQ (2000) Neural networks for classification: a survey. IEEE Trans Syst Man Cybern. Part C Appl Rev 30(4):451–462

Chapter 13
MARS Use in Evaluating Entry-Type Excavation Stability

The mining industry relies heavily on the use of empirical charts for entry-type excavation design and assessment of stability. The commonly used empirical design method, called the critical span graph based on an extensive case history database of cut and fill mining operations in Canada, was specifically developed for rock stability analysis in entry-type excavations. This span design chart plots the critical span against rock mass rating for the observed case histories and has been widely accepted by many mining operations for an initial span design of cut and fill stopes. Other methods, either based on classical regression and classification statistical techniques or even the supervised machine learning methods, have also been proposed to classify the observed cases into stable, potentially unstable, and unstable groups. The main purpose of this chapter is to present the use of MARS_LR method proposed in Chap. 12 in evaluating the stability of underground entry-type excavations.

13.1 Background

Entry-type mining methods, such as cut and fill, have been replaced in many mining operations by much lower cost, non-entry mining methods. In many mines, however, cut and fill is still appropriate in conditions where the hanging wall is very weak or ore body contacts are quite irregular. Therefore, entry-type mining methods are still desirable. In addition, in view of the relatively high costs associated with cut and fill mining, there can be significant savings from an improved, more reliable and safe, back stability design (Wang et al. 2000).

There are many empirical and numerical methods for assessing the back stability in entry-type underground openings. It is generally difficult to obtain reliable input data describing the rock mass conditions for numerical design approaches, for which the mechanical parameters are usually derived from rock mass classification systems such as the tunneling quality index Q system, rock mass rating RMR system, and the geological strength index GSI system, through empirical equations. Furthermore,

© Science Press and Springer Nature Singapore Pte Ltd. 2019
W. Zhang, *MARS Applications in Geotechnical Engineering Systems*,
https://doi.org/10.1007/978-981-13-7422-7_13

the numerical approach is more time-consuming to obtain reliable results. In a production environment, a quick and reliable design method is needed that can give the field engineer or technician safe guidelines for opening dimensions. Based on the collection of field data and an assessment of stability, usually, these techniques can only reliably be used in conditions similar to the conditions under which the empirical data sets were collected.

The most widely used empirical design method called the "critical span graph" was developed by Lang (1994) (University of British Columbia) to provide a practical design tool developed specifically for spans in entry-type excavations. It is based on an extensive case history database of cut and fill mining in Canada and defines stable, potentially unstable, and unstable cases in span areas on a graph of rock mass rating (Bieniawski 1976) (RMR_{76} rock mass performance parameter) against the span between pillars. This graph has been accepted by many mining operations for an initial span design of cut and fill stopes, and it enables an operator to assess the stability of mine openings with respect to a rock mass. Recently, García-Gonzalo et al. (2016) adopted the supervised machine learning classifiers (support vector machine and extreme learning machines) to define stability areas of the critical span graph. Although the predictive capacities of these two models are satisfactory, they have been criticized for the computational inefficiency and the poor model interpretability.

13.2 Critical Span Graph and the Database

The critical span graph developed by Lang (1994) was specifically used for spans in entry-type excavations, by compiling and plotting the 172 observations of a database from entry-type case histories, on a span versus RMR_{76} graph, to enable future prediction of stable spans given the RMR_{76} of the stope (García-Gonzalo et al. 2016). Rock mass rating (RMR_{76}) is widely accepted and used as a rock mass classification system. It combines the most significant geomechanical parameters and represents them with an overall comprehensive index of rock mass quality. To apply this system, the rock mass is divided into a number of structural domains, and each is assessed. Weighted ratings were used considering that the parameters are not equally important. The summation of all parameters produces an RMR value ranging from 0 to 100. Upon this approach, a description of the rock mass based on classes can be defined by its number (for example, RMR < 20 is a very poor rock, and RMR > 80 is a very good rock). The advantage of this system is that only a few basic parameters that are related to the geometry and mechanical conditions of the rock mass are required. Its popularity stemmed from the fact that it could be used for excavation design in rock with significant capacity to predict the excavation stand-up time (García-Gonzalo et al. 2016). Considering that the RMR classification system has been updated several times since its first publication, it is generally referred to with a subscript indicating the year to identify the version of the classification being used. As an example of the RMR_{76} application in characterizing the rock mass, the mean values of the different geologic parameters and the resulting values of rock mass quality index are shown

Table 13.1 RMR$_{76}$ at Detour Lake Mine (adapted from García-Gonzalo et al. 2016)

Category	Main zone		Talc zone	
	Description	Rating	Description	Rating
Strength	160–180 MPa	13	35–50 MPa	4
RQD	90%	17	80%	16
Joint spacing	0.4 m	16	0.3 m	9
Joint condition	Smooth, hard, tight	17	Smooth surfaces, soft	10
Groundwater	None	10	None	10
Joint orientation		0		0
Total RMR$_{76}$		73		49

Fig. 13.1 Critical span graph (adapted from Lang 1994)

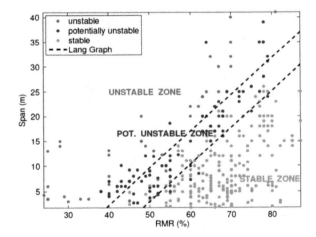

in Table 13.1 in the two main areas of the operation of the Detour Lake Gold Mine (Lang 1994), where the values of the original database compiled by Lang (1994) were obtained.

The critical span graph developed by Lang (1994) consists of two straight lines that divide the RMR$_{76}$ versus span graph into three zones (stable, potentially unstable and unstable rock), as shown in Fig. 13.1.

In 2002, the database was expanded to 292 observations by a further study conducted by Wang et al. (2002), with the addition of case histories from six more mining operations and using a neural network analysis for the construction of the stability graph, and updating the critical span graph. Subsequently, in 2003, Kumar (2003) incorporated 107 more new observations for a final database with 399 cases and updated the critical span graph also using a neural network analysis. Although some contradictions in the successive enlargements of the database made by researchers at British Columbia University (Lang 1994; Wang et al. 2002; Kumar 2003) are

Table 13.2 Data sources of the case histories (adapted from Kumar 2003)

Mines	Cases	Stable (S)	Potentially unstable (P)	Unstable (U)
Detour Lake Mine	172	94	37	41
Detour Lake Mine	22	10	0	12
Photo Lake Mine	6	0	6	0
Olympias Mine	13	4	1	8
Brunswick Mining	17	5	3	9
Musslewhite Mine	46	35	10	1
Snip Mine	16	12	2	2
Red Lake Mine	107	81	19	7
Summary	399	241	78	80

observed, the database established in the work of Kumar (2003), which is the one that incorporates the largest number of cases, was taken as the reference for conducting the present study. This final database consisted of stope behavior data from eight operating mines in Canada with observational data from 399 operational case histories. The data case history sources are shown in Table 13.2. Each case history contains information about rock mass conditions expressed as an RMR_{76} value, span, and the corresponding rock stability.

In this database, the RMR_{76} ranges from 24 to 87 and the span from 2 to 41 m. The RMR_{76} values for 57% of the cases were concentrated in the range of 60–80. The span values from 3 to 30 m constitute 95% of the cases. The input data were obtained from different mines that had different personnel surveying the stope dimensions and estimating the RMR_{76}. This brought in huge variability or inaccuracy into the input data. However, the span estimation error should be substantially below 1 m, which is within the tolerance of the graphical design approach. The variability in estimating the RMR_{76} value can be more significant and will depend on the level of experience of the engineer conducting the classification work. The critical span graph and its updates have been widely accepted in the mining industry and provide a quick and simple tool to estimate a maximum span that may be designed based on the observed RMR_{76} value.

Almost all previous work with the critical span graph classifies the data into three groups, because field observations are grouped into the stable, unstable, and potentially unstable categories. García-Gonzalo et al. (2016) considered an alternative construction of the critical span graph, requiring only the information of field observations corresponding to the stable and unstable classes, using a probabilistic classification that allows one to define soft boundaries between the two classes considered. As these two classes are easier to assess by the engineer, the error due to incorrect assessment can thus be minimized. This chapter uses the MARS_LR approach to probabilistically define the soft boundaries between the stable and unstable cases.

As listed in Table 13.2, the database complied by Kumar (2003) consists of 399 cases, 241 stable cases, 78 potentially unstable cases, and 80 unstable observations.

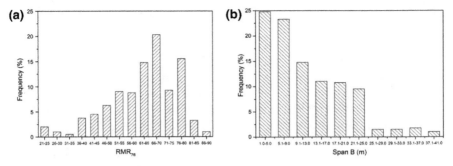

Fig. 13.2 Histograms of: **a** RMR$_{76}$, and **b** span B

Figure 13.2 plots the histograms of the two inputs RMR$_{76}$ (increment of 5) and span (increment of 4 m). For the details of these cases, the reader is referred to Kumar (2003).

13.3 The Developed MARS_LR Model and Modeling Results

To build the MARS_LR model, the target output called stability potential SP for the 399 cases are represented by three numbers: 0 denotes stable, 0.5 stands for potentially unstable, while 1 is for unstable. It should be noted that Juang et al. (2013a, b) have adopted a similar liquefaction potential concept to assess the probability of liquefaction occurrence. The modeling result of SP ranging from 0 to 0.2 is regarded as stable, while SP ranging from 0.8 to 1.0 is regarded as unstable. The SP values ranging from 0.2 to 0.8 are considered to be potentially unstable. Of the 399 data sets, 299 were randomly selected as the training patterns, while the remaining 100 were used for testing purpose. The criterion of data pattern selection was based on ensuring that the statistical properties including the mean and standard deviations of the training and testing subsets were similar to each other.

By trial and error, the estimation of stability potential SP using the MARS_LR model with second-order interaction adopted eight BFs of linear spline function. The predictions are shown in Fig. 13.3 for the training and testing patterns. For the training patterns, of the 182 stable cases, nine cases are regarded as potentially unstable and ten cases are estimated as unstable. Of the 76 potentially unstable cases, six cases are regarded as stable while most cases are estimated as potentially unstable and unstable. Of the 41 unstable cases, only one case is assessed as stable. For the testing patterns, of the 58 stable cases, three cases are regarded as potentially unstable and three cases are estimated as unstable. Of the 27 potentially unstable cases, six cases are regarded as stable, while most cases are estimated as potentially unstable and unstable. Of the 15 unstable cases, none is assessed as stable.

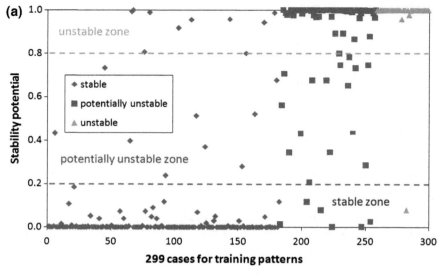

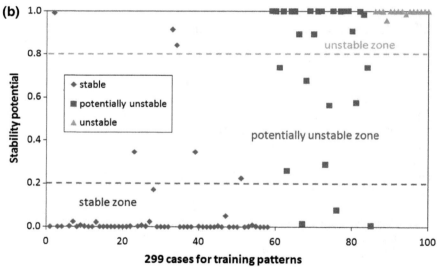

Fig. 13.3 Modeling results of: **a** training data; **b** testing data

Table 13.3 Confusion matrix for the training and testing results of MARS_LR

Confusion matrix	Training patterns		Testing patterns		Overall	
	Stable	Unstable	Stable	Unstable	Stable	Unstable
Stable	163 (89.6%)	19 (10.4%)	52 (89.7%)	6 (10.3%)	215 (89.6%)	25 (10.4%)
Unstable	7 (6.0%)	110 (94.0%)	6 (14.3%)	36 (85.7%)	13 (8.2%)	146 (91.8%)

Table 13.3 lists the confusion matrix for the training, testing, and the overall patterns. It is obvious that the Type I error is very low (8.2%). In addition, the developed MARS_LR model has an overall success rate of 89.6% in estimating stable cases and overall success rate of 91.8% in estimating unstable cases. The modeling results are satisfactory in terms of accuracy, compared with the model accuracy of 82% by the support vector machine method by García-Gonzalo et al. (2016) with classification into three groups and the model accuracy of 98% by support vector machine method with a binary probabilistic classification. It should also be noted that the model accuracy of 88% is obtained from the extreme learning machine with the classification of the data sets into three groups by García-Gonzalo et al. (2016). The modeling results of both the MARS_LR method and the learning classifiers presented an improved design method in terms of accuracy over the original design method by Lang (1994) and the neural network approach by Wang et al. (2000).

13.4 Model Interpretability

Table 13.4 lists the BFs and their corresponding equation for the developed MARS_LR model. It is observed from Table 13.4 that interactions have occurred between BFs (four of the eight BFs are interaction terms). The presence of interactions suggests that the built MARS model is not simply additive and that interactions play a significant role in building an accurate model for stability potential predictions. This again indicates that MARS is capable of capturing the nonlinear and complex relationships without making any specific assumption about the underlying functional relationship between the input variables and the dependent response. In addition, it can be observed that there are five knots for variable RMR, at values of RMR = 38, 55, 58, 66, and 70 while there are two knots for variable span B, at values of $B = 11$, and 20 m. These knot values enable engineers to have an insight and understanding of where significant changes in the data may occur. Table 13.5 also shows another advantage of the developed MARS_LR model, i.e., the good interpretability over the machine learning classifiers by García-Gonzalo et al. (2016) and the neural network by Wang et al. (2000).

Table 13.4 BFs and MARS_LR expression

BFs	Expression	BFs	Expression
BF1	max(0, RMR − 70)	BF5	BF2 × max(0, RMR − 58)
BF2	max(0, 20 − B)	BF6	BF1 × max(0, 20 − B)
BF3	max(0, RMR − 38) × max(0, 20 − B)	BF7	max(0, B − 11)
BF4	max(0, RMR − 55)	BF8	BF7 × max(0, 66 − RMR)

$$f(x) = \frac{1}{(1+e^{-(20.6+1.69\times BF1-0.686\times BF2-0.049\times BF3-2.27\times BF4+0.21\times BF5-0.112\times BF6+1.11\times BF7-0.117\times BF8))})}$$

13.5 Summary

This chapter has demonstrated the use of the MARS_LR approach in evaluating the stability of underground entry-type excavations based on an extensive case history database of cut and fill mining in Canada. Comparisons indicate that the MARS_LR performs as well as, or better than, the machine learning classifiers and the neural network approach in terms of accuracy. However, considering its simplicity of interpretation, predictive accuracy, its data-driven, and adaptive nature, its ability to map the interaction between variables and the relatively low number of Type I error predictions, the use of MARS_LR model in evaluating the stability of underground entry-type excavations is promising.

It should be noted that as the proposed MARS_LR model was developed using database records in Canada, it should be used with caution in other countries. In addition, more reliable case records should be included to expend the database, for which the use of a Bayesian updating scheme is promising, as has used in braced excavations by Wang et al. (2012a, b). Furthermore, the main limitation of this empirical design approach is the quality of RMR. It must be recognized, as with all empirical design approaches, that the limitation of the predictive solution is largely governed by the data collected and how the data is processed.

References

Bieniawski ZT (1976) Rock mass classification in rock engineering. In: Bieniawski Z Ied) Exploration for rock engineering, vol 1. Balkema, Cape Town, South Africa, pp 97–106

García-Gonzalo E, Fernández-Muñiz Z, Nieto PJG, Sánchez AB, Fernández MM (2016) Hard-rock stability analysis for span design in entry-type excavations with learning classifiers. Materials 9:531

Juang CH, Luo Z, Atamturktur S, Huang H (2013a) Bayesian updating of soil parameters for braced excavations using field observations. J Geotech Geoenviron Eng 139:395–406

Juang CH, Ching J, Wang L, Khoshnevisan S, Ku CS (2013) Simplified procedure for estimation of liquefaction-induced settlement and site-specific probabilistic settlement exceedance curve using cone penetration test (CPT). Can Geotech J 50(10):1055–1066

Kumar P (2003) Development of empirical and numerical design techniques in burst prone ground at the red lake mine. Master's thesis, University of British Columbia, Vancouver, BC, Canada

Lang B (1994) Span design for entry-type excavations. Master's thesis, University of British Columbia, Vancouver, BC, Canada

Wang J, Pakalnis R, Milne D, Lang B (2000) Empirical underground entry-type excavation span design modification. In: Montreal 2000: 53th Annual Conference of the Canadian Geotechnical Society

Wang J, Milne D, Pakalnis R (2002) Application of a neural network in the empirical design of underground excavation spans. Min Technol IMM Trans Sec A 111:73–81

Wang DD, Qiu GQ, Xie WB, Wang Y (2012a) Deformation prediction model of surrounding rock based on GA-LSSVM-markov. Nat Sci 4(2):85–90

Wang L, Ravichandran N, Juang CH (2012b) Bayesian updating of KJHH model for prediction of maximum ground settlement in braced excavations using centrifuge data. Comput Geotech 44:1–8

Chapter 14
Conclusions and Recommendations

14.1 Summary and Conclusions

This book has five objectives. The first one is to introduce a promising MARS procedure for numerical mapping. The second is to show the main advantages of MARS
over other methods including the variety of neural networks, the extreme learning
machine ELM, the linear genetic programming LGP, multi-expression programming
MEP, standard genetic programming GP, and adaptive neuro-fuzzy inference system
ANFIS, for complex data mapping in high-dimensional data. The third is to present
some applications of MARS algorithm in big data geotechnical problems, such as the
HP pile drivability analysis. The fourth is to demonstrate the procedures of MARS
use, including the model development, model interpretation, and parametric sensitivity analysis. The last is to illustrate the modified MARS procedure for pattern
recognition/classification (MARS_LR), such as the liquefaction assessment and the
stability evaluation of the entry-type excavation.

14.1.1 Main Advantages of MARS

MARS is capable of modeling the nonlinear relationships involving a multitude of
variables with interactions among each other without making any specific assumption
about the underlying functional relationship between the input variables and the
response. The MARS approach is also computationally efficient and is able to provide
the relative importance of the input variables. Since MARS explicitly defines the
intervals for the input variables, the model enables engineers to have an insight and
understanding of where significant changes in the data may occur.

© Science Press and Springer Nature Singapore Pte Ltd. 2019
W. Zhang, *MARS Applications in Geotechnical Engineering Systems*,
https://doi.org/10.1007/978-981-13-7422-7_14

14.1.2   Applications of MARS Algorithm in Big Data Geotechnical Problems

A database containing 4072 pile data sets with a total of seventeen variables including the pile parameters, the design variables from hammer and its cushion materials, and soil parameters is adopted to verify the MARS use for big data drivability predictions in relation to the prediction of the maximum compressive stresses, maximum tensile stresses, and blow per foot. Performance measures indicate that the MARS algorithm for analyses of pile drivability is promising, with high accuracy, less computation time, and easy to interpret models.

14.1.3   MARS_LR for Pattern Recognition/Classification

MARS_LR performs as well as, or better than, the machine learning classifiers and the neural network approach in terms of accuracy. However, considering its simplicity of interpretation, predictive accuracy, its data-driven and adaptive nature, its ability to map the interaction between variables, and the relatively low number of Type I error predictions, the use of MARS_LR model in stability evaluation of underground entry-type excavations and liquefaction assessment is promising.

14.2   Recommendations for Future Work

Some suggestions for future research are given below:

(1) Various self-pruning neural network programs can be considered to alleviate the problem of finding the optimal architecture (Williams 1995; Burden and Winkler 2009). These programs automatically add and delete hidden neurons as they learn the mapping function. For the MARS algorithm adopted in this study, the number of basis functions and the choices of basis functions and the knots are determined by generalized cross-validation. Further work may also consider other machine learning approaches such as the genetic algorithms and support vector machines to optimize the knot locations and the basis functions.

(2) On top of the optimization of the knot locations and the basis functions, the MARS algorithm may also be amended/modified for reliability analysis, i.e., the combined use of MARS and reliability approaches like FORM, PEM, and MCS. However, since the built MARS models adopt the piece-wise linear segment to approximate the real curve/surface relationship, its combined use with FORM is still doubted. In this regard, the MARS and MCS methods can be combined since the latter is based on the samplings technique, which ignores the shape of the line or surface representing the limit state functions.

(3) More complicated geotechnical engineering problems with several million or more data sets and several hundred input parameters at hand, i.e., the bored tunnels or the mass rapid transit system excavation, should be utilized to test the MARS use. Its successful application would pave a sound foundation for the development of unmanned tunneling technique. Also in this regard, a dynamic decision-making procedure based on the MARS algorithm is indispensable.

References

Burden FR, Winkler DA (2009) An optimal self-pruning neural network and nonlinear descriptor selection in QSAR. QSAR Comb Sci 28:1092–1097

Williams PM (1995) Bayesian regularization and pruning using a Laplace prior. Neural Comput 7:117–143

Appendix A
MATLAB BP Algorithm Adopted for the BPNN Models

```
clear
clc                    % clear the command window


rand('seed', 1);
% randn('seed', 1);


   hiddenminno =10
   best=10
   for run=1:1:best
        maxepoch =500;


% trainset;
Variable1=[…];
Variable2=[…];
Dependent=[…];
traininput1=[Variable1; Variable2];
traintarget1= Dependent;
```

© Science Press and Springer Nature Singapore Pte Ltd. 2019
W. Zhang, *MARS Applications in Geotechnical Engineering Systems*,
https://doi.org/10.1007/978-981-13-7422-7

```matlab
% normalize
[traininput, mintraininput1, maxtraininput1, traintarget, mintraintarget1, maxtraintarget1] =
premnmx(traininput1, traintarget1);


% testset;
testVariable1=[...];
testVariable2=[...];
testDependent =[...];
testinput1=[testVariable1; testVariable2];
[testinput] = tramnmx(testinput1,mintraininput1,maxtraininput1);

 testtarget= testDependent;
[PN,minp,maxp,TN,mint,maxt] = premnmx(testinput1,testtarget);
[attrno,trainexpno] = size(traininput);      % 'attrno' is the number of attributes,
'trainexpno' is the number of training examples

% rand('state',sum(100*clock));
    % train the neural networks
    net = newff(MinMax(traininput),[hiddenminno 1],{'logsig' 'tansig'});
    net.trainParam.epochs = maxepoch;
    net.trainParam.goal = 1e-10;
    net.trainParam.max_fail=100;
    net.trainParam.min_grad=1e-15;
    net.trainParam.mu_max=1e20;
    net.trainParam.mu_dec=0.7;
    net.trainParam.mu_inc=1.03;
    net.trainParam.lr=0.01;
net = train(net,traininput,traintarget);


    % test
[n,testexpno] = size(testinput);              % 'testexpno' is the number of test
examples, 'n' is useless
[m,trexpno] = size(traininput);
 output2 = zeros(1,trexpno);
 output1 = zeros(1,testexpno);
    output1 = sim(net,testinput);
        [output_te] = postmnmx(output1,mintraintarget1,maxtraintarget1);
    output2 = sim(net,traininput);
        [output_tr] = postmnmx(output2,mintraintarget1,maxtraintarget1);
```

```
    figure(run);
    [m,b,r]=postreg(output_te,testtarget);
      mse_te = mse(output_te - testtarget);                    % obtain the mean squared error of
  the ensemble
      mse_tr = mse(output_tr - traintarget1);
      mse_te1 = mse(output1 - TN) ;                            % according to the scaled value
      mse_tr2 = mse(output2 - traintarget)  ;
      fprintf('single hidden=%g run=%g r=%-12.5g mse_te=%-12.5g mse_tr=%-12.5g',
  hiddenminno, run, r, mse_te, mse_tr);
      fprintf('\n\n');
    end
```

% end of function

For the mathematical explanations of the MATLAB built-in functions *premnmx*, *tramnmx*, *postmnmx*, *postreg*, …, please refer to MATLAB Help on Neural Network Toolbox.

After obtaining the optimal BPNN model, to obtain the weights connecting the input layer and the hidden layer, type "w1=net.iw{1,1}" in the MATLAB command window; to obtain the bias values at neurons of the hidden layer, type "theta1=net.b{1}" in the command window; to get the weights connecting the hidden layer and the output layer, type "w2=net.lw{2,1}" in the command window; and to obtain the bias value at the output layer, type "theta2=net.b{2}" in the command window.

Appendix B
MARS Algorithms Adopted for the Built Models Using MATLAB

```
clc
clear
cd 'C:\ASCEBP'
a=load('pillaraverage.txt');
X1=a(:,1);
X2=a(:,2);
X=[X1, X2];
Y=a(:,3);
params = aresparams(4, [], false, [], [], 2);

model = aresbuild(X, Y, params)

aresanova(model, X, Y)

Yq = arespredict(model, X);

mu=mean(Y);

J=sum((Yq-Y).^2);

S=sum((Y-mu).^2);

tmse=sum((Y-Yq).^2);

mse=tmse/length(Y)

rmse=sqrt(mse);

r2=1-J/S

b=load('pillaraveragetesting.txt');

Xt(:,1)=b(:, 1);

Xt(:,2)=b(:, 2);

Yt=b(:, 3);

[MSE, RMSE, RRMSE, R2] = arestest(model, Xt, Yt)

aresplot(model)
```

© Science Press and Springer Nature Singapore Pte Ltd. 2019
W. Zhang, *MARS Applications in Geotechnical Engineering Systems*,
https://doi.org/10.1007/978-981-13-7422-7

aresparams

```
function trainParams = aresparams(maxFuncs, c, cubic, cubicFastLevel, ...selfInteractions,
maxInteractions, threshold, prune, useMinSpan, ...useEndSpan, maxFinalFuncs)


if (nargin < 1) || (isempty(maxFuncs))
    trainParams.maxFuncs = 21;
else
    trainParams.maxFuncs = maxFuncs;
end
if (nargin < 2) || (isempty(c))
    trainParams.c = 3;
else
    trainParams.c = c;
end


if (nargin < 3) || (isempty(cubic))
    trainParams.cubic = true;
else
    trainParams.cubic = cubic;
end

if (nargin < 4) || (isempty(cubicFastLevel))
    trainParams.cubicFastLevel = 2;
else
    trainParams.cubicFastLevel = cubicFastLevel;
end
if (nargin < 5) || (isempty(selfInteractions))
    trainParams.selfInteractions = 1;
else
    trainParams.selfInteractions = selfInteractions;
end

if (trainParams.cubic) && (trainParams.selfInteractions > 1)
    trainParams.selfInteractions = 1;
    disp('Warning: trainParams.selfInteractions value reverted to 1 due to piecewise-cubic
model setting.');
end
```

```matlab
if (nargin < 6) || (isempty(maxInteractions))
   trainParams.maxInteractions=1;
else
   trainParams.maxInteractions = maxInteractions;
end


if (nargin < 7) || (isempty(threshold))
   trainParams.threshold = 1e-4;
else
   trainParams.threshold = threshold;
end
if (nargin < 8) || (isempty(prune))
   trainParams.prune = true;
else
   trainParams.prune = prune;
end


if (nargin < 9) || (isempty(useMinSpan))
   trainParams.useMinSpan = -1; % default = -1 = automatic
else
   if useMinSpan == 0
      trainParams.useMinSpan = 1; % 1 and 0 is the same here (no endspan)
   else
      trainParams.useMinSpan = useMinSpan;
   end
end


if (nargin < 10) || (isempty(useEndSpan))
   trainParams.useEndSpan = -1;
else
   if useEndSpan == 0
      trainParams.useEndSpan = 1;
   else
      trainParams.useEndSpan = useEndSpan;
   end
end
```

```
if (nargin < 11) || (isempty(maxFinalFuncs))
   trainParams.maxFinalFuncs = Inf;
else
   trainParams.maxFinalFuncs = maxFinalFuncs;
end

return
```

aresbuild ————————————————————————

```
function [model, time] = aresbuild(Xtr, Ytr, trainParams, weights, modelOld, verbose)

if nargin < 2
   error('Too few input arguments.');
end
if (isempty(Xtr)) || (isempty(Ytr))
   error('Training data is empty.');
end
if (~isfloat(Xtr)) || (~isfloat(Ytr))
   error('Data type should be floating-point.');
end
[n, d] = size(Xtr); % number of data cases and number of input variables
if size(Ytr,1) ~= n
   error('The number of rows in the matrix and the vector should be equal.');
end
if size(Ytr,2) ~= 1
   error('Ytr should have one column.');
end

if (nargin < 3) || (isempty(trainParams))
   trainParams = aresparams();
end
if trainParams.maxInteractions >= 2
   trainParams_actual_c = trainParams.c;
else
   trainParams_actual_c = 2*trainParams.c/3; % penalty coefficient for additive modeling
end
```

```matlab
if (trainParams.cubic) && (trainParams.selfInteractions > 1)
    trainParams.selfInteractions = 1;
    disp('Warning: trainParams.selfInteractions value reverted to 1 due to piecewise-cubic
model setting.');
end
if trainParams.cubic

    doCubicFastLevel = trainParams.cubicFastLevel;
    if trainParams.cubicFastLevel > 0
        trainParams.cubic = false;
    end
else
    doCubicFastLevel = -1;
end

if trainParams.useMinSpan == 0
    trainParams.useMinSpan = 1;
end

if trainParams.useEndSpan == 0
    trainParams.useEndSpan = 1;
end
if (nargin < 4)
    weights = [];
else

    if (~isempty(weights)) && ...
        ((size(weights,1) ~= n) || (size(weights,2) ~= 1))
        error('weights vector is of wrong size.');
    end
end
wd = diag(weights);

if nargin < 5
    modelOld = [];
end

if (nargin < 6) || (isempty(verbose))
    verbose = true;
end
```

```matlab
if verbose, fprintf('Building ARES model...\n'); end
ws = warning('off');
tic;


maxIters = floor(trainParams.maxFuncs / 2);
YtrMean = mean(Ytr);
YtrSS = sum((Ytr - YtrMean) .^ 2);
minX = min(Xtr);
maxX = max(Xtr);


if trainParams.useEndSpan < 0
    endSpan = getEndSpan(d); % automatic
else
    endSpan = trainParams.useEndSpan;
end


if isempty(modelOld)
    X = ones(n,1);
    err = 1; % normalized error for the constant model
    model.coefs = YtrMean;
    model.knotdims = {};
    model.knotsites = {};
    model.knotdirs = {};
    model.parents = [];
    model.trainParams = [];
    model.MSE = Inf;
    model.GCV = Inf;
else
    model = modelOld; % modelOld is the initial model
end


if endSpan*2 >= n
    if isempty(modelOld)
        model.MSE = YtrSS / n;
        model.GCV = gcv(model, model.MSE, n, trainParams_actual_c);
        if trainParams.cubic
            model.t1 = [];
            model.t2 = [];
        end
    end
else
```

```
% FORWARD PHASE

if isempty(modelOld) % no forward phase when modelOld is used

  if verbose, fprintf('Forward phase .'); end


  % create sorted lists of data cases for knot placements
  [sortedXtr sortedXtrInd] = sort(Xtr);
  if trainParams.useEndSpan ~= 1  % throw away data cases at the ends of the intervals
    sortedXtr = sortedXtr(endSpan:end-(endSpan-1),:);
    sortedXtrInd = sortedXtrInd(endSpan:end-(endSpan-1),:);
  end

  if trainParams.cubic
    tmp_t1 = [];
    tmp_t2 = [];
  end
  basisFunctionList = []; % will contain candidate basis functions
  numNewFuncs = 0; % how many basis functions added in the last iteration


  % the main loop of the forward phase
  for depth = 1 : maxIters
    basisFunctionList = createList(basisFunctionList, Xtr, sortedXtr, sortedXtrInd, ...
                        n, d, model, numNewFuncs, trainParams, endSpan);


    % stop the forward phase if basisFunctionList is empty
    if isempty(basisFunctionList)
      if trainParams.cubic
        t1 = tmp_t1;
        t2 = tmp_t2;
      end
      break;
    end

    tmpErr = inf(1,size(basisFunctionList,2));
    tmpCoefs = inf(length(model.coefs)+2, size(basisFunctionList,2));
    Xtmp = zeros(n,size(X,2)+2);
    if ~trainParams.cubic
      Xtmp(:,1:end-2) = X;
    end
```

```
% try all the reflected pairs in the list
for i = 1 : size(basisFunctionList,2)
    if trainParams.cubic
        [t1 t2 dif] = findsideknots(model, basisFunctionList{1,i}, basisFunctionList{2,i},
..., minX, maxX, tmp_t1, tmp_t2);
        Xtmp(:,1:end-2) = X;
        % update basis functions with the updated side knots
        for j = 1 : length(model.knotdims)
            if dif(j)
                Xtmp(:,j+1) = createbasisfunction(Xtr, Xtmp, model.knotdims{j},
model.knotsites{j}, ... model.knotdirs{j}, model.parents(j), minX, maxX, t1(j,:), t2(j,:));
            end
        end
        % New basis function
        dirs = basisFunctionList{3,i};
        Xtmp(:,end-1) = createbasisfunction(Xtr, Xtmp, basisFunctionList{1,i},
basisFunctionList{2,i}, ... dirs, basisFunctionList{4,i}, minX, maxX, t1(end,:), t2(end,:));
        if isnan(Xtmp(1,end-1)), Xtmp(:,end-1) = []; end
        % Reflected partner
        dirs(end) = -dirs(end);
        Xtmp(:,end) = createbasisfunction(Xtr, Xtmp, basisFunctionList{1,i},
basisFunctionList{2,i}, ... dirs, basisFunctionList{4,i}, minX, maxX, t1(end,:), t2(end,:));
        if isnan(Xtmp(1,end)), Xtmp(:,end) = []; end
    else
        % New basis function
        dirs = basisFunctionList{3,i};
        Xtmp(:,end-1) = createbasisfunction(Xtr, Xtmp, basisFunctionList{1,i}, ...
                    basisFunctionList{2,i}, dirs, basisFunctionList{4,i}, minX, maxX);
        if isnan(Xtmp(1,end-1)), Xtmp(:,end-1) = []; end
        % Reflected partner
        dirs(end) = -dirs(end);
        Xtmp(:,end) = createbasisfunction(Xtr, Xtmp, basisFunctionList{1,i}, ...
                    basisFunctionList{2,i}, dirs, basisFunctionList{4,i}, minX, maxX);
        if isnan(Xtmp(1,end)), Xtmp(:,end) = []; end
    end
```

```
            if size(Xtmp,2) == size(X,2)+2 % both basis functions created
                [coefs tmpErr(i)] = lreg(Xtmp, Ytr, weights, wd);
                tmpErr(i) = tmpErr(i) / YtrSS;
                tmpCoefs(:,i) = coefs;
            elseif size(Xtmp,2) == size(X,2)+1 % one of the basis functions not created
                [coefs tmpErr(i)] = lreg(Xtmp, Ytr, weights, wd);
                tmpErr(i) = tmpErr(i) / YtrSS;
                tmpCoefs(:,i) = [coefs; NaN];
                Xtmp = [Xtmp zeros(n,1)];
            else
                tmpErr(i) = Inf;
                Xtmp = [Xtmp zeros(n,2)];
            end
        end


        [newErr, ind] = min(tmpErr); % find out the best modification


        % stop the forward phase if no correct model was created or if the decrease in error is
below the threshold
        if (isnan(newErr)) || (err(end) - newErr < trainParams.threshold)
            if trainParams.cubic
                t1 = tmp_t1;
                t2 = tmp_t2;
            end
            break;
        end


        if trainParams.cubic
            [t1 t2 dif]=findsideknots(model,basisFunctionList{1,ind},
basisFunctionList{2,ind}, ... d, minX, maxX, tmp_t1, tmp_t2);
            % update basis functions with the updated side knots
            for j = 1 : length(model.knotdims)
                if dif(j)
                    X(:,j+1)=createbasisfunction(Xtr,X,model.knotdims{j}, model.knotsites{j}, ...
model.knotdirs{j}, model.parents(j), minX, maxX, t1(j,:), t2(j,:));
                end
            end
            % Add the new basis function
            dirs = basisFunctionList{3, ind};
```

```
        Xn=createbasisfunction(Xtr,X,basisFunctionList{1,ind}, basisFunctionList{2,ind},
...dirs,basisFunctionList{4,ind},minX, maxX, t1(end,:), t2(end,:));
        if isnan(Xn(1)), Xn = []; end
        % Add the reflected partner
        dirs(end) = -dirs(end);
        Xn2=createbasisfunction(Xtr,X,basisFunctionList{1,ind},
basisFunctionList{2,ind},...dirs,basisFunctionList{4,ind}, minX, maxX, t1(end,:), t2(end,:));
        if isnan(Xn2(1)), Xn2 = []; end
        X = [X Xn Xn2];
        if ~isempty(Xn) && ~isempty(Xn2) % one of the basis functions is not created
            t1(end+1,:) = t1(end,:);
            t2(end+1,:) = t2(end,:);
        end
    else
        dirs = basisFunctionList{3, ind};
        % Add the new basis function
        Xn = createbasisfunction(Xtr, X, basisFunctionList{1,ind}, ...
            basisFunctionList{2,ind}, dirs, basisFunctionList{4,ind}, minX, maxX);
        if isnan(Xn(1)), Xn = []; end
        % Add the reflected partner
        dirs(end) = -dirs(end);
        Xn2 = createbasisfunction(Xtr, X, basisFunctionList{1,ind}, ...
            basisFunctionList{2,ind}, dirs, basisFunctionList{4,ind}, minX, maxX);
        if isnan(Xn2(1)), Xn2 = []; end
        X = [X Xn Xn2];
    end
    model.coefs = tmpCoefs(:,ind);


    % add the basis functions to the model
    numNewFuncs = 0;
    dirs = basisFunctionList{3, ind};
    if ~isempty(Xn)
        model.knotdims{end+1,1} = basisFunctionList{1, ind};
        model.knotsites{end+1,1} = basisFunctionList{2, ind};
        model.knotdirs{end+1,1} = dirs;
        model.parents(end+1,1) = basisFunctionList{4, ind};
        numNewFuncs = numNewFuncs + 1;
    else
        model.coefs(end) = [];
    end
    if ~isempty(Xn2)
        dirs(end) = -dirs(end);
        model.knotdims{end+1,1} = basisFunctionList{1, ind};
        model.knotsites{end+1,1} = basisFunctionList{2, ind};
```

```
      model.knotdirs{end+1,1} = dirs;
      model.parents(end+1,1) = basisFunctionList{4, ind};
      numNewFuncs = numNewFuncs + 1;
    else
      model.coefs(end) = [];
    end


    if verbose, fprintf('..'); end
    err(end+1) = newErr;
    if (newErr < trainParams.threshold) || ...
      (length(model.coefs) + 2 > n)
        break;
    end


    if trainParams.cubic
      tmp_t1 = t1;
      tmp_t2 = t2;
    end
    basisFunctionList(:,ind) = [];
  end % end of the main loop

    if verbose, fprintf('\n'); end


end % end of "isempty(modelOld)"


if isempty(modelOld)
  if (doCubicFastLevel == 1) || ...
    ((doCubicFastLevel >= 2) && (~trainParams.prune))
      % turn the cubic modelling on
      trainParams.cubic = true;
      [t1 t2] = findsideknots(model, [], [], d, minX, maxX, [], []);
      % update all the basis functions
      for i = 1 : length(model.knotdims)
          X(:,i+1) = createbasisfunction(Xtr, X, model.knotdims{i}, model.knotsites{i}, ...
                  model.knotdirs{i}, model.parents(i), minX, maxX, t1(i,:), t2(i,:));
      end
      [model.coefs model.MSE] = lreg(X, Ytr, weights, wd);
      model.MSE = model.MSE / n;
  else
      model.MSE = err(end) * YtrSS / n;
  end
```

```matlab
      model.GCV = gcv(model, model.MSE, n, trainParams_actual_c);
      if trainParams.cubic
        model.t1 = t1;
        model.t2 = t2;
      end
    end


% BACKWARD PHASE


  if trainParams.prune


    if verbose, fprintf('Backward phase .'); end


    if ~isempty(modelOld) % create basis functions from scratch when modelOld is used
      if (doCubicFastLevel == -1) || (doCubicFastLevel >= 2)
      % create all the basis functions (linear) from scratch
      X = ones(n,length(model.knotdims)+1);
      for i = 1 : length(model.knotdims)
        X(:,i+1)=createbasisfunction(Xtr, X, model.knotdims{i}, model.knotsites{i}, ...
            model.knotdirs{i}, model.parents(i), minX, maxX);
      end
      [model.coefs model.MSE] = lreg(X, Ytr, weights, wd);
      model.MSE = model.MSE / n;
      model.GCV = gcv(model, model.MSE, n, trainParams_actual_c);
    else
      % create all the basis functions (cubic) from scratch
      t1 = model.t1;
      t2 = model.t2;
      X = ones(n,length(model.knotdims)+1);
      for i = 1 : length(model.knotdims)
        X(:,i+1)=createbasisfunction(Xtr, X, model.knotdims{i}, model.knotsites{i}, ...
            model.knotdirs{i}, model.parents(i), minX, maxX, t1(i,:), t2(i,:));
      end
    end
  end
```

```
    models = {model};
    mses = model.MSE;
    gcvs = model.GCV;


    % the main loop of the backward phase
    for j = 1 : length(model.knotdims)
        tmpErr = inf(1, length(model.knotdims));
        tmpCoefs = inf(length(model.coefs)-1, length(model.knotdims));


    % try to delete model's basis functions one at a time
    for k = 1 : length(model.knotdims)
        Xtmp = X;
        Xtmp(:,k+1) = [];
        if trainParams.cubic
            % create temporary t1, t2, and model with a deleted basis function

            tmp_t1 = t1;
            tmp_t1(k,:) = [];
            tmp_t2 = t2;
            tmp_t2(k,:) = [];
            tmp_model.knotdims = model.knotdims;
            tmp_model.knotdims(k) = [];
            tmp_model.knotsites = model.knotsites;
            tmp_model.knotsites(k) = [];
            tmp_model.knotdirs = model.knotdirs;
            tmp_model.knotdirs(k) = [];
            tmp_model.parents = model.parents;
            tmp_model.parents(k) = [];
            tmp_model.parents = updateParents(tmp_model.parents, k);
            [tmp_t1 tmp_t2 dif] = findsideknots(tmp_model, [], [], d, minX, maxX, tmp_t1,
tmp_t2);

            % update basis functions with the updated side knots
            for i = 1 : length(tmp_model.knotdims)
                if dif(i)

                    Xtmp(:,i+1) = createbasisfunction(Xtr, Xtmp, tmp_model.knotdims{i},
tmp_model.knotsites{i}, ...tmp_model.knotdirs{i}, tmp_model.parents(i), minX, maxX,
tmp_t1(i,:), tmp_t2(i,:));
                end
```

```
      end
    end
    [coefs tmpErr(k)] = lreg(Xtmp, Ytr, weights, wd);

    tmpCoefs(:,k) = coefs;

  end


  [dummy, ind] = min(tmpErr); % find out the best modification
  X(:,ind+1) = [];
  model.coefs = tmpCoefs(:,ind);
  model.knotdims(ind) = [];
  model.knotsites(ind) = [];
  model.knotdirs(ind) = [];
  model.parents(ind) = [];
  model.parents = updateParents(model.parents, ind);
  if trainParams.cubic
    t1(ind,:) = [];
    t2(ind,:) = [];
    [t1 t2 dif] = findsideknots(model, [], [], d, minX, maxX, t1, t2);
    % update basis functions with the updated side knots
    for i = 1 : length(model.knotdims)
      if dif(i)
        X(:,i+1)=createbasisfunction(Xtr,X,model.knotdims{i}, odel.knotsites{i}, ...
            model.knotdirs{i}, model.parents(i), minX, maxX, t1(i,:), t2(i,:));
      end
    end
    model.t1 = t1;
    model.t2 = t2;
  end


  models{end+1} = model;
  mses(end+1) = tmpErr(ind) / n;
  gcvs(end+1) = gcv(model, mses(end), n, trainParams_actual_c);


  if verbose, fprintf('.'); end
end % end of the main loop
```

```
    if trainParams.maxFinalFuncs >= length(models{1}.coefs)
       [g, ind] = min(gcvs);
    elseif trainParams.maxFinalFuncs > 1
       [g, ind] = min(gcvs(end-trainParams.maxFinalFuncs+1:end));
       ind = ind + length(gcvs) - trainParams.maxFinalFuncs;
    else
       g = gcvs(end);
       ind = length(gcvs);
    end
    model = models{ind};


    if doCubicFastLevel >= 2
       % turn the cubic modeling on
       trainParams.cubic = true;
       [t1 t2] = findsideknots(model, [], [], d, minX, maxX, [], []);
       % update all the basis functions
       X = ones(n,length(model.coefs));
       for i = 1 : length(model.knotdims)
          X(:,i+1) = createbasisfunction(Xtr, X, model.knotdims{i}, model.knotsites{i}, ...
                model.knotdirs{i}, model.parents(i), minX, maxX, t1(i,:), t2(i,:));
       end
       model.t1 = t1;
       model.t2 = t2;
       [model.coefs model.MSE] = lreg(X, Ytr, weights, wd);
       model.MSE = model.MSE / n;
       model.GCV = gcv(model, model.MSE, n, trainParams_actual_c);
    else
       model.MSE = mses(ind);
       model.GCV = g;
    end


    if verbose, fprintf('\n'); end


  end % end of "trainParams.prune"


end % end of "if endSpan*2 >= n"
```

```
    model.trainParams = trainParams;
    model.minX = minX;
    model.maxX = maxX;
    model.endSpan = endSpan;


    time = toc;
    if verbose
        fprintf('Number of basis functions in the final model: %d\n', length(model.coefs));
        fprintf('Total effective number of parameters: %0.1f\n', ...
            length(model.coefs) + model.trainParams.c * length(model.knotdims) / 2);
        maxDeg = 0;
        if length(model.knotdims) > 0
            for i = 1 : length(model.knotdims)
                if length(model.knotdims{i}) > maxDeg
                    maxDeg = length(model.knotdims{i});
                end
            end
        end
        fprintf('Highest degree of interactions in the final model: %d\n', maxDeg);
        fprintf('Execution time: %0.2f seconds\n', time);
    end
    warning(ws);
    return
```

aresanova ──────────────────────────────────

```
function aresanova(model, Xtr, Ytr)


if nargin ~= 3
    error('The number of input arguments should be axactly three.');
end
if (isempty(Xtr)) || (isempty(Ytr))
    error('Data is empty.');
end
```

```matlab
n = size(Xtr,1);
if size(Ytr,1) ~= n
    error('The number of rows in the matrix and the vector should be equal.');
end
if size(Ytr,2) ~= 1
    error('Ytr should have one column.');
end


if model.trainParams.cubic
    fprintf('Type: piecewise-cubic\n');
else
    fprintf('Type: piecewise-linear\n');
end
fprintf('GCV: %0.3f\n', model.GCV);
fprintf('Total number of basis functions: %d\n', length(model.coefs));
fprintf('Total effective number of parameters: %0.1f\n', ...
        length(model.coefs) + model.trainParams.c * length(model.knotdims) / 2);
fprintf('ANOVA decomposition:\n');
fprintf('Func.\t\tSTD\t\t\tGCV\t\t#basis\t#params\t\tvariable(s)\n');
counter = 0;
for i = 1 : model.trainParams.maxInteractions
    combs = nchoosek(1:length(model.minX),i);
    for j = 1 : size(combs,1)
        [modelReduced usedBasis] = aresanovareduce(model, combs(j,:), true);
        if length(usedBasis) > 0
            counter = counter + 1;
            fprintf('%d\t\t', counter);
            fprintf('%7.3f\t', std(arespredict(modelReduced, Xtr))); % standard deviation of the
ANOVA function
            modelReduced = deleteBasis(model, usedBasis);
            Yq = arespredict(modelReduced, Xtr);
            MSE = mean((Ytr - Yq).^2);
            fprintf('%11.3f\t\t', gcv(modelReduced, MSE, n, model.trainParams.c)); % GCV when
the basis functions are deleted
            fprintf('%6d\t', length(usedBasis)); % the number of basis functions for that ANOVA
function
```

```matlab
        fprintf('%7.1f\t\t', length(usedBasis) + model.trainParams.c * length(usedBasis) / 2);
% effective parameters
        fprintf('%d ', combs(j,:)); % used variables
        fprintf('\n');
    end
  end
end
return


function modelReduced = deleteBasis(model, nums)
modelReduced = model;
for i = length(nums) : -1 : 1
  modelReduced.coefs(nums(i)+1) = [];
  modelReduced.knotdims(nums(i)) = [];
  modelReduced.knotsites(nums(i)) = [];
  modelReduced.knotdirs(nums(i)) = [];
  modelReduced.parents(nums(i)) = [];
  if modelReduced.trainParams.cubic
    modelReduced.t1(nums(i),:) = [];
    modelReduced.t2(nums(i),:) = [];
  end
end
modelReduced.parents(:) = 0;
if modelReduced.trainParams.cubic
  % correct the side knots for the modified cubic model
  [modelReduced.t1 modelReduced.t2] = ...
  findsideknots(modelReduced, [], [], size(modelReduced.t1,2), modelReduced.minX,
modelReduced.maxX, [], []);
end
return


function g = gcv(model, MSE, n, c)
% Calculates GCV from model complexity, its Mean Squared Error, number of data cases n,
and penalty coefficient c.
enp = length(model.coefs) + c * length(model.knotdims) / 2; % model's effective number of
parameters
if enp >= n
  g = Inf;
```

```
else
    p = 1 - enp / n;
    g = MSE / (p * p);
end
return
```

arespredict ————————————————————————————

```
function Yq = arespredict(model, Xq)
if nargin < 2
    error('Too few input arguments.');
end
X = ones(size(Xq,1),length(model.knotdims)+1);
if model.trainParams.cubic
    for i = 1 : length(model.knotdims)
        X(:,i+1) = createbasisfunction(Xq, X, model.knotdims{i}, model.knotsites{i}, ...
                model.knotdirs{i}, model.parents(i), model.minX, model.maxX, model.t1(i,:),
model.t2(i,:));
    end
else
    for i = 1 : length(model.knotdims)
        X(:,i+1) = createbasisfunction(Xq, X, model.knotdims{i}, model.knotsites{i}, ...
                model.knotdirs{i}, model.parents(i), model.minX, model.maxX);
    end
end
Yq = X * model.coefs;
return
```

arestest ————————————————————————————

```
function [MSE, RMSE, RRMSE, R2] = arestest(model, Xtst, Ytst)


if nargin < 3
    error('Too few input arguments.');
end
if (isempty(Xtst)) || (isempty(Ytst))
    error('Data is empty.');
end
```

```
if (size(Xtst, 1) ~= size(Ytst, 1))
    error('The number of rows in the matrix and the vector should be equal.');
end
if size(Ytst,2) ~= 1
    error('The vector Ytst should have one column.');
end
MSE = mean((arespredict(model, Xtst) - Ytst) .^ 2);
RMSE = sqrt(MSE);
if size(Ytst, 1) > 1
    RRMSE = RMSE / std(Ytst, 1);
    R2 = 1 - MSE / var(Ytst, 1);
else
    RRMSE = Inf;
    R2 = Inf;
end
return
```

aresplot ————————————————————————————

```
function aresplot(model, minX, maxX, vals, gridSize)
if nargin < 1
    error('Too few input arguments.');
end
if (nargin < 2) || (isempty(minX))
    minX = model.minX;
else
    if length(minX) ~= length(model.minX)
        error('Vector minX is of wrong size.');
    end
end
if (nargin < 3) || (isempty(maxX))
    maxX = model.maxX;
else
    if length(maxX) ~= length(model.maxX)
        error('Vector maxX is of wrong size.');
    end
end
if (nargin < 5) || (isempty(gridSize))
    gridSize = 50;
end
```

```
if length(model.minX) < 2
    step = (maxX - minX) / gridSize;
    X = [minX:step:maxX]';
    plot(X, arespredict(model, X));
    return
end


if (nargin < 4) || (isempty(vals))
    ind1 = 1; vals(1) = NaN;
    ind2 = 2; vals(2) = NaN;
    if length(minX) > 2
        for i = 3 : length(minX)
            vals(i) = (maxX(i) - minX(i)) / 2;
        end
    end
else
    if length(minX) ~= length(vals)
        error('Vector vals is of wrong size.');
    end
    tmp = 0;
    for i = 1 : length(vals)
        if isnan(vals(i))
            if tmp == 0
                ind1 = i;
                tmp = 1;
            elseif tmp == 1
                ind2 = i;
                tmp = 2;
            else
                tmp = 3;
                break;
            end
        end
    end
    if tmp ~= 2
        error('Vector vals should contain NaN exactly two times.');
    end
end
```

```
step1 = (maxX(ind1) - minX(ind1)) / gridSize;
step2 = (maxX(ind2) - minX(ind2)) / gridSize;


[X1,X2] = meshgrid(minX(ind1):step1:maxX(ind1), minX(ind2):step2:maxX(ind2));
XX1 = reshape(X1, numel(X1), 1);
XX2 = reshape(X2, numel(X2), 1);


X = zeros(size(XX1,1), length(minX));
X(:,ind1) = XX1;
X(:,ind2) = XX2;
for i = 1 : length(minX)
   if (i ~= ind1) && (i ~= ind2)
      X(:,i) = vals(i);
   end
end


YY = arespredict(model, X);
Y = reshape(YY, size(X1,1), size(X2,2));
surfc(X1, X2, Y);
return
```

```
else
    p = 1 - enp / n;
    g = MSE / (p * p);
end
return
```

arespredict ————————————————————————————————

```
function Yq = arespredict(model, Xq)
if nargin < 2
    error('Too few input arguments.');
end
X = ones(size(Xq,1),length(model.knotdims)+1);
if model.trainParams.cubic
    for i = 1 : length(model.knotdims)
        X(:,i+1) = createbasisfunction(Xq, X, model.knotdims{i}, model.knotsites{i}, ...
                model.knotdirs{i}, model.parents(i), model.minX, model.maxX, model.t1(i,:),
model.t2(i,:)));
    end
else
    for i = 1 : length(model.knotdims)
        X(:,i+1) = createbasisfunction(Xq, X, model.knotdims{i}, model.knotsites{i}, ...
                model.knotdirs{i}, model.parents(i), model.minX, model.maxX);
    end
end

Yq = X * model.coefs;
return
```

arestest ————————————————————————————————

```
function [MSE, RMSE, RRMSE, R2] = arestest(model, Xtst, Ytst)


if nargin < 3
    error('Too few input arguments.');
end
if (isempty(Xtst)) || (isempty(Ytst))
    error('Data is empty.');
end
```

```matlab
if (size(Xtst, 1) ~= size(Ytst, 1))
    error('The number of rows in the matrix and the vector should be equal.');
end
if size(Ytst,2) ~= 1
    error('The vector Ytst should have one column.');
end
MSE = mean((arespredict(model, Xtst) - Ytst) .^ 2);
RMSE = sqrt(MSE);
if size(Ytst, 1) > 1
    RRMSE = RMSE / std(Ytst, 1);
    R2 = 1 - MSE / var(Ytst, 1);
else
    RRMSE = Inf;
    R2 = Inf;
end
return
```

aresplot ————————————————————————————

```matlab
function aresplot(model, minX, maxX, vals, gridSize)
if nargin < 1
    error('Too few input arguments.');
end
if (nargin < 2) || (isempty(minX))
    minX = model.minX;
else
    if length(minX) ~= length(model.minX)
        error('Vector minX is of wrong size.');
    end
end
if (nargin < 3) || (isempty(maxX))
    maxX = model.maxX;
else
    if length(maxX) ~= length(model.maxX)
        error('Vector maxX is of wrong size.');
    end
end
if (nargin < 5) || (isempty(gridSize))
    gridSize = 50;
end
```